True Nature

A Theory of
Sexual Attraction

PERSPECTIVES IN SEXUALITY
Behavior, Research, and Therapy

Series Editor: RICHARD GREEN

University of Cambridge
Cambridge, England, United Kingdom and
Gender Identity Clinic
Charing Cross Hospital
London, England, United Kingdom

True Nature
A Theory of Sexual Attraction

Michael R. Kauth
Mental Health Service
Veterans Affairs Medical Center
and
Department of Psychiatry
Louisiana State University School of Medicine
New Orleans, Louisiana

Kluwer Academic / Plenum Publishers
New York, Boston, Dordrecht, London, Moscow

Library of Congress Cataloging-in-Publication Data

Kauth, Michael R.
    True nature: a theory of sexual attraction/Michael R. Kauth.
       p. cm. —(Perspectives in sexuality)
    Includes bibliographical references and index.
    ISBN 0-306-46390-3
      1. Sexual attraction. I. Title. II. Series.

   HO23 .K317 2000
   306.7—dc21
                                 00-024105

ISBN 0-306-46390-3

©2000 Kluwer Academic/Plenum Publishers
233 Spring Street, New York, New York 10013

http://www.wkap.nl/

10 9 8 7 6 5 4 3 2 1

A C.I.P. record for this book is available from the Library of Congress

Printed in the United States of America

*To those who are willing to examine
their most cherished beliefs*

Preface

I have long been awe-struck by authors' claims that their books had been in the making for 5, or 10, or even 15 years. I now have a better appreciation of the work involved in bringing a book to press.

The seeds of this project have had a long germination. The impetus for this book began more than 10 years ago when I was a graduate student in clinical psychology. Having an interest in human sexuality—and in theories on the forms of sexual attraction specifically—I was perplexed by various perspectives on this subject. Disciplines of thought that I encountered— medicine, evolutionary biology, developmental psychology, gay/lesbian theory, social constructionism, anthropology, Marxism, Christianity, and others—perceived the issue so differently, so strongly, with almost no overlap. I was fascinated that the question of how and why one is attracted to either one or both sexes could elicit such conviction and divergent points of view. There seemed to be no easy way to resolve these differences. Still, what frustrated me most in my readings were several conceptual problems among the two prominent proponents of contemporary sexuality theory— scientists and social constructionists.

One of my first frustrations with biomedical and social scientists who write about sexuality was that they often define sexual attraction in strict behavioral terms, as completed observable sexual acts—observable in the sense that such acts or their consequences are seen by others. Marriage, same-sex erotic behavior, erection, and orgasm are examples of observable acts that have been used to infer human sexual orientation. Such a strategy, of course, has severe limitations. Being private, human sexual behavior may never accurately be observed, particularly if the behavior is stigmatized. However, more to the point, sexual behavior is not a good indicator of human sexual desire or motivation.

A second disappointment was that biomedical writers frequently describe their human subjects as somehow existing apart from their social environment. By focusing exclusively on biological causes of sexual attraction, biomedical researchers appear not to consider that culture influences the expression and perception of sexual feelings and behavior—

even more, that definitions of what is "sexual" or even "homosexual" differ across historical periods and cultures. How biology and culture *interact* is almost never discussed. If mentioned at all, culture may be evoked to explain away unwelcome variation in sexual behavior and erotic feelings.

Finally, in the past, scientists have seemed unaware that their own assumptions, their line of inquiry, and their conclusions about research were influenced by the social context in which they lived and worked. Social constructionists enjoy pointing out this fact. For example, consistent with current Western social beliefs, biomedical writers have generally assumed that sexual attraction comes in two exclusive forms—homosexual and heterosexual. Bisexuality as a persistent form of attraction is ignored, although at least male bisexuality been common throughout history. Biomedical writers rarely conceive of sexual attraction on a continuum or as a proportion of gendered desires. How social beliefs, expectations, and experience modify biological dispositions to produce sexual behavior is largely unexplored by biomedical theorists.

When I read social constructionist literature on sexuality, additional frustrations surfaced. For the most part, social constructionists purport their positions on sexual attraction development by refuting biomedical theories. However, refuting one set of theories does not and should not lead automatically to an uncritical acceptance of an alternate explanation. Uniformly, constructionsts have failed to apply to their own ideas the same intense scrutiny that they have directed toward biomedical theories. The latest example of this lopsided "debate" comes from Edward Stein (1999) in *The Mismeasure of Desire*. Stein, who usually challenges constructionist ideas, argues that scientific efforts to study same-sex attraction have been imprecise and biased and should perhaps not be attempted at all. Stein's solution is to accept constructionist explanations about sexuality without giving them the same critical examination that he gave biomedical theories.

In addition, I was exasperated with social constructionist writers for generally conceptualizing people as devoid of organic bodies and somehow independent of biology. Physiology, neurochemistry, sex hormones, sex drive, and evolution are treated as largely irrelevant to the development of various forms of sexual attraction and to sexuality. Social constructionist writers usually dismiss attribution of physiological influences as an attempt to medicalize sex and sexuality.

And lastly, I found that social constructionist writers present an uncharitable and misleading view of Science. Although the sins of Science have been great and numerous—which constructionists are fond of noting—there is no recognition from them that Science is a self-correcting process founded on collections of probable facts. No doubt, the inherent tentativeness of scientific knowledge is most evident in the study of internal and external human events and personal identity as they relate to sexuality. Yet, social constructionist writers seem to imply all too often that tentative knowledge is trivial and that lack of absolute certainty equals absolute falsehood or unbelievability.   If that were so, Science would be useless and could accomplish nothing.

In short, my dissatisfaction with current critical thinking on sexuality led me to explore and compare a variety of theories on attraction. My own exploration was framed within the developmental perspective of my professional training—that human beings are a species with a genetic history and a psychology; that they are neurochemical learning machines who create sophisticated social cultures and in turn are significantly affected by them. These are the assumptions on which my thinking about sexuality is based.

My intent in this book is to provide a rational and probable explanation for the development of varied forms of sexual attraction and their persistence. However, a good practical theory, in my opinion, is a beginning, not an end. As I noted earlier, Science is a process of steps and missteps—a self-correcting process guided by a critical methodology. I hope that my explanation of human sexual attraction raises many questions that stimulate thought and spark a *true* debate among sexuality theorists. I hope that my explanation lures theorists who are in entrenched conceptual positions to step outside of their fortifications for a moment. I hope that this book will bring sexologists another step closer to actualizing a practical interactive theory of human sexual attraction.

Before we begin, a few final qualifications are necessary: the ideas and opinions proffered in this volume are mine and do not represent the views or positions of the Veterans Health Administration or the United States government. This book is not a product of service with or financial remuneration by the United States government. All errors, inaccuracies, unintended misrepresentations, and inarticulate narration are mine as well, and I bear sole responsibility for them.

Acknowledgments

Many people contributed to the making of this book. First, I must give credit to the numerous sexuality theorists and researchers whose works I have devoured over the years. The works of these men and women not only inspired me but also provided countless hours of intellectual enjoyment. I must also thank Dan Landis who fed my desire to write a text on human sexuality and showed me that it was possible. Patrick Hopkins and David Powell helped me greatly by challenging my early ideas about sexuality. Their critiques and suggestions for this manuscript were invaluable.

Second, I must thank my two library assistants, James Podboy and Malissa Parker, as well as a host of helpful librarians who spent hours searching for obscure materials and making copies.

Thanks to Peggy Schlegel and Elizabeth Knoll, who read primitive versions of my manuscript and provided constructive comments. Many thanks also to my dear friend, David Gochman, whose incisive editorial comments and critique helped to further refine the final version of this manuscript.

I extend a special acknowledgment to Mike Ross, who contributed an early draft of what later developed into chapter 2. I am especially grateful for Mike's continued encouragement and sage advice during the Dark Days of this project. A special thanks also to Richard Green for seeing potential in my work and to Mariclaire Cloutier and Tuan Hoang at Kluwer Academic/Plenum Publishers for their editorial assistance in bringing this manuscript to fruition.

Lastly, I wish to acknowledge my parents, Richard and Viola, for their unwavering pride in me and to thank Matthew Horsfield for his companionship and patience during the tedious, time-consuming development of this volume.

Without all of you, this project would still be a collection of undeveloped ideas and scattered sheaves of notes.

Contents

True Nature

A Theory of
Sexual Attraction

CHAPTER 1: OBFUSCATION AND CLARIFICATION: AN INTRODUCTION

"Human nature" does not describe people. — Ruth Hubbard, *The Political Nature of "Human Nature"*

Human kind cannot bear very much reality. — T. S. Eliot, *Four Quartets*

I am curious. I want to know things. In particular, I am curious about things related to sex, like erotic attraction and sexual identity. You are probably just as curious about these things as I am. That's why we are meeting here—in this book.

I have a lot of questions about sexual attraction: Why are some men and women sexually attracted only to men? Why are other men and women sexually attracted only to women, and still other people to both sexes? While we assume that the function of other-sex (male-female) attraction is to facilitate finding a mate and reproduction, what benefit, if any, might same-sex erotic attraction have for survival? How in fact has same-sex erotic attraction survived over generations? Is exclusive other-sex erotic attraction a reproductive advantage over sexual attraction to both sexes, or is it the other way around? Is sexual attraction, whether to one or both sexes, actually a by-product of something else?

In the past, were people in Western cultures and non-Western cultures who were attracted to individuals of the same sex similar or different from "lesbians" and "gay" men today? Was the sex of the person to whom one was attracted significant or irrelevant? If the partner's sex was unimportant, what characteristics were important in structuring romantic relationships? Clearly, we are not attracted to all members of a sex. How is it then that some personality, behavioral or physical characteristics seem important to sexual attraction and others do not? How was sexual attraction—as opposed to sexual activity with a marriage partner—manifest in ancient and non-Western cultures? What influence does culture—and, by *culture*, I broadly mean the

1

social institutions, customs, beliefs, politics, and technology connected in time to a specific group of people—have on the expression of sexual attraction? How has contemporary Western culture influenced sexual attraction? How have negative attitudes about same-sex erotic attraction shaped sexual behavior and social identity? Why do ordinary people, as well as scientists, seem to be less interested in the origins of other-sex erotic attraction than in discovering the "secret" of homosexuality? Is same-sex erotic attraction really so much more mysterious and unique than other-sex attraction? Whatever the answer is, how then do beliefs about same-sex erotic attraction influence the expression of other-sex attraction? Do all cultures share the same fascination with and, yet, fear of same-sex erotic attraction as is apparent in popular Western culture?

And, what role does biology have in determining the sex to which we are erotically attracted? What is the physiology and neurochemistry of sexual attraction? How separate are sexual attraction, sexual desire, sex, and gender? If genetics or hormones influence sexual attraction, can sexual attraction ever change from one sex to the other? Is sexual attraction immutable, or fluid, or something in between?

Why do *I* have the particular sexual and erotic feelings that I experience? Why do I not have other erotic feelings? Would my sexual feelings and direction of sexual attraction be different if I lived in another culture? Does sexual attraction develop similarly in all people? Could there be different developmental pathways for other-sex erotic attraction than for same-sex erotic attraction? Is there a continuum of sexual orientation— composed of various sexual attractions or sexualities—rather than two or three discrete types?

During the past 20 years of reading and thinking about sexual attraction, I have discovered that everyone—and I mean *everyone*—has an opinion about the cause and purpose of sexual attraction. What is more, these beliefs are quite personal and emotionally defended. This book will probably do little to influence people who are emotionally or politically invested in a particular theory of sexual attraction. This discussion is not an emotional appeal for the "rightness" of my view. Rather, readers who are willing to acknowledge that they do not have absolute answers to questions about their own or another's sexual orientation will find the discussion in this book interesting, informative, and provocative. Through the framework of science, my 20-year interdisciplinary search has led to a working hypothesis about the development and function of general human sexual attraction—and many additional questions. There are, of course, limitations to a scientific approach to sexual attraction.

Science cannot claim unquestionable Truth, particularly when applied to human social behavior. Science is based on systematic observation and predictability. Numerous and variable factors of influence within evolving sets of conditions make careful observation and interpretation about the origin and purpose of sexual attraction extremely difficult. Scientists are left with, at best, probable causes of sexual attraction that are subject to modification when more reliable observations are available. Exact cause and unwavering certainty are elusive in Science. People who are looking for absolute Truth and certainty about sexual attraction will not find it in a scientific discourse; such people are more likely to find comforting simple explanations for sexuality in the clearly (although not always logically) delineated ontology of religion. The crux of the problem may be the difference between a search for meaning—that is, the purpose for or *why* something happens—and an understanding of sequence—that is, *how* a thing occurs.

The value or meaning of why something happens is nearly impossible for Science to answer; questions about how a thing takes place, however, are possible to answer with a scientific approach. The solution to the problem is in recognizing the limits of what Science can do and the implicit assumptions of our questions. Often when people ask explanatory how questions, they imply or conflate their desire for answers about meaning or why something happens. This is not a surprising mistake. For many of us, explanations about the sequence of events, while important, may not be especially satisfying if we are looking for answers to personal questions about sexual attraction or anticipate answers that will challenge or support long-held personal beliefs about it. The questions we ask determine what answers we can find. Let me be clear about the questions I am asking: I want to know the process by which kinds of sexual attraction develop—that is, the "how" of sexual attraction—and I also wish to know *why* varieties of sexual attraction exist, a less probable but potentially more satisfying answer.

How and why questions produce *proximate* and *ultimate* explanations for a phenomena, respectively (Symons, 1979/1987). According to evolutionary biologist Donald Symons, *proximate* causes are immediate or close (to the organism) factors that describe in what way the organism came to develop and exhibit the behavioral patterns of interest. He notes that proximate causes involve a "complement of genes and particular environmental consequences in the presence of particular stimuli," whereas *ultimate* causes—*why* explanations—are related to "the behavior of ancestors to promote reproductive success by exhibiting behavior in similar circumstances" (p. 5). Ultimate causes are not necessarily innate or unlearned. But, because our ancestors are long dead, knowing the ultimate cause of the varieties of human sexual behavior is nearly impossible. At best, we can propose logical, consistent, falsifiable hypotheses about why, for example, some people have exclusive same-sex erotic feelings or why people value particular characteristics over others in a sexual mate. Evolutionary biologists are interested in ultimate explanations of human sexual behavior. Although this book considers ultimate causes for human sexual attraction, it is not strictly a book on evolutionary theory or ultimate causation.

On the other hand, psychologists, sociologists, neurobiologists, and social constructionists (including gay, queer, and gender theorists) are more interested in proximate causes—that is, what is going on in the individual's brain and particular social environment that produces these types of sexual attraction. Proximalists are a diverse group. Generally, proximalists view ultimate explanations as assumed, speculative, or beside the point, and evolutionary biologists such as Symons (1979/1987) think that proximate explanations are wholly incomplete. Neither side gives much consideration to the other kind of causation. However, several proximate causes will be discussed in this book, although the presence of certain genes, hormones, brain sites, social experiences, and social roles are not the final word in the development of sexual attraction.

Ultimate and proximate causes of sexual attraction are important interrelated questions. Unfortunately, most theorists of human sexual attraction favor one kind of explanation and reject or ignore the other. This practice has limited theories of sexual attraction to narrow descriptions of distant evolutionary processes that bear little resemblance to contemporary human life or to accounts of individual sexual behavior and physiology that seem disembodied and free-floating in the history of human behavior. Symons's *The Evolution of Human Sexuality* (1979/1987) is an example of the former, and John Money's *Gay, Straight, and In-Between* (1988) is an example of the latter.

What is more, many proximal theorists have restricted their causal hypotheses to linear, discipline-specific accounts, giving only lipservice to the interaction between biology and the social environment. Biomedical writers such as Simon LeVay, for example, often limit their discussion to specific neurohormonal events or the relative size of brain structures without addressing how particular environmental events or learning experiences or the timing of those phenomena contribute to differences in sexual attraction. Linear reductionist descriptions of size differences in brain structures, although informative, tell us nothing about the context in which those differences developed. (This problem is discussed in more detail in chapter 5). Similarly, social constructionists who write about sexual attraction— Kenneth Plummer and Carol Vance, to name two—generally accept the idea of a biologically grounded sexual desire but avoid talking about how sexual desire shapes cultural beliefs and politics and is in turn shaped by them. (This issue is taken up in chapter 6).

Interactionist theories of sexual attraction—theories that explain sexual attraction as the interaction between biological processes and social experiences—are few and provide mainly proximal explanations. The interactionist theories that exist focus mostly on particular causes, either psychosocial or biomedical. John Money (1988), for example, describes the development of (homo)sexual attraction as the interaction between fetal neurohormones and early postnatal social experiences. (I compare Money's model to my own in chapter 5). Money's account is primarily a theory of how events "go wrong" to produce homosexuality. Donald Symons (1979/1987) also proposes that human sexual attraction is the interactive product of biology and the social environment and concludes that hormonal imbalances produce same-sex erotic attraction. The implication, again, is that same-sex erotic attraction is always a mistake and that other-sex attraction never is. My theory by contrast describes varied sexual attractions—exclusive attraction to one sex and attraction to both sexes—as a nonpathological, developmental process and is not as such a theory of same-sex erotic attraction.

Interactionist theories of sexual attraction are not new. More than one hundred years ago sexologist Havelock Ellis proposed an interactionist explanation for same-sex erotic attraction—known then as *sexual inversion*—in his *Studies in the Psychology of Sex*, the six volumes of which were published between 1897 and 1910. In Volume I, entitled *Sexual Inversion*, Ellis (1936) asserted that inversion is a "favorable organic predisposition" (p. 324) that requires an "exciting cause." He then went on to describe several social factors such as same-sex erotic experiences at school and failed romantic relationships that facilitated the process of inversion. However, 100 years of theorizing has witnessed little advancement in the specificity of the interaction between biology and the social environment that contributes to the development of human sexual attraction. Narrow discipline-specific theories, rather than interdisciplinary ones, have driven theorizing about sexual attraction. The next step is an *interdisciplinary interactionist theory* of sexual attraction that accounts for the interlocking sequence of biological and psychosocial developmental events, as well as the powerful influence of cultural beliefs, social practices, and institutions. Interactionist theories of sexual attraction must also address the role of social constructionism, an idea that has dominated the study of sexuality for the past 20 years. Although radical social constructionism rejects all scientific explanations for sexual attraction, I believe it is possible to meld middle ground constructionist ideas with an empirical approach and gain a dynamic theory of sexual attraction.

Interactionist theories, although practical and sensible, have never been popular. Thus, what I present here is a book about the development and function of human sexual attraction that in all likelihood will make academic disciplines and theoretic purists unhappy. This book attempts to integrate disparate data from several investigational fields into a holistic, synthetic account of the development of various types of sexual attraction. The idea that sexual attraction is the result of the interaction between biology and the social environment is not new, but I hope to provide a fresh account of how these elements fit together. I also view the social environment in a broad context, as something more than a collection of interacting individuals (Symons, 1979/1987); *social environment* includes cultural ideas, religion, history, customs, and social institutions. Furthermore, I offer a hypothesis for the adaptation of varied sexual attraction, such as same-sex eroticism, and how that adaptation can be recognized today.

Let me say again that what I present is not a book about homosexuality, although data and observations about same-sex erotic attraction make up a large portion of this text. Because most sexuality research has focused on explaining homosexuality, it is inevitable that any general theory of human sexual attraction must discuss this history and utilize its rich database. However, at this time, forming a theory of same-sex erotic attraction, even within the larger framework of a general theory of sexual attraction, raises ethical questions. Udo Schüklenk and Michael Ristow (1996) warned recently that research into the causes of same-sex erotic attraction is probably unethical, because knowledge about same-sex attraction has traditionally been used to oppress "lesbians" and "gay" men.[1] Therefore, the authors declare that investigations for the pursuit of knowledge should be viewed with suspicion. Schüklenk and Ristow claim that at heart research regarding the origin of same-sex erotic attraction is a search for a cure for homosexuality—only because homosexuality is perceived as a social problem are researchers even interested in it. Schüklenk and Ristow further point out that objective knowledge about "gay" men and "lesbians" has not appreciably changed the antigay attitudes of many people, including social scientists; therefore, research into the cause of same-sex attraction for the sake of knowledge is in effect feeding oppression. Only in an unbiased society, could research about the causes of same-sex erotic attraction be "ethically justified" in their view (Schüklenk & Ristow, 1996, 25).

Unfortunately, we live in a prejudiced society where "lesbians" and "gay" men are harmed by scientific data; we live in a society in which many people are not treated with respect, dignity, or fairness. *Homonegativism*—negative beliefs about homosexuals—is pervasive in Western and many non-Western cultures. And, there is always a risk that information will be misinterpreted or deliberately misused by individuals and groups who hold negative attitudes about "lesbians" and "gay" men. However, I resist Schüklenk and Ristow's chilling and simplistic conclusion that research into the origins of homosexuality is unethical. Not all research about the cause of same-sex erotic attraction implies the objective to cure homosexuality, and the absence of an ideal society is not a sufficient reason to avoid research about same-sex erotic attraction. Although the motives of antigay researchers such as endocrinologist Gunter Dörner are obvious and unmistakable, it is a far and bizarre stretch to conclude that neurobiologist Simon LeVay, geneticist Dean Hamer, and psychologist Michael Bailey—all of whom view same-sex erotic attraction as a normal variation—secretly wish to cure "lesbians" and "gay" men. The objective of this book, for example, is the understanding of the origins and development of various forms of sexual attraction and is not about "curing" any particular form of attraction. Variety of forms is not in itself evidence of pathology; Nature's rule is diversity, not homogeneity. What is more, the desire to understand a phenomenon is not synonymous with wanting to change it.

Scientists have a responsibility to pursue knowledge and to monitor how their assumptions color their conclusions; the peer review process and discussion of findings via journals and professional conferences are two ways that scientists attempt to correct for biased conclusions. Of course, scientists and nonscientists alike have cherished and defended the most heinous, inhumane beliefs with moral and objective-sounding arguments. Yet, to conclude, as do Schüklenk and Ristow, that research in some areas should be restricted because information can be and has been used to oppress or punish the subjects of study, means that research into many areas of social controversy would be prohibited. Scientists would have to stop studying people who are subject to stigmatization such as ethnic minorities, single mothers, women, the severely mentally ill, the disabled, runaways, the homeless, substance abusers, and those who hold politically undesirable beliefs because irresponsible people and powerful social groups could misuse these data. Science is then constrained to the study of noncontroversial, nonpolitical topics. Schüklenk and Ristow misunderstand the process and objectives of Science; research is never unbiased, and the purpose of scientific knowledge is not to change social attitudes, persuade, or to ensure that marginalized people are treated fairly, although these are admirable and desirable goals. Science is a tool to gain knowledge. As scientists and members of society, it is our responsibility to use that knowledge fairly and wisely.

As I stated earlier, I am interested in a general theory of sexual attraction that addresses both the *how* and *why* questions of causation. It is probably true that had I lived in a culture that was less sexually focused and that did not stigmatize same-sex erotic attraction I might have little interest in this topic. My culture has shaped my desire for certain kinds of knowledge. However, my motive is not to find a cure for same-sex erotic attraction but to understand better all types of sexual attraction. My goal is not to change social attitudes or persuade society to be more tolerant; I do not have the means to accomplish those tasks. At best, I hope to offer some possible answers, stimulate thought, identify hidden assumptions, and challenge those assumptions about sexual attraction.

1.1 CHALLENGING ASSUMPTIONS

Particularly hazardous assumptions are associations with the terms *natural* and *normal*. Either overtly or covertly, these concepts invariably pop up in discussions about sexuality. In general, but especially within scientific discourse, one must be careful about how words are used. These particular terms have multiple meanings and yet are often employed to assert moral values and to suggest what ought or ought not to be with regard to sexual attraction and behavior. Since *natural* and *normal* are so frequently associated with sexual attraction and so often misapplied, a brief discussion about these terms is warranted.

Schüklenk and Ristow (1996) in their critique of research on homosexuality complained rightly that writers often treat *natural* and *normal* as synonyms and *prescriptive rules*; in other words, what is *natural* or *normal* is both good and what ought to be. Said in a different way, *natural* and *normal* sexuality <u>should</u> be supported and encouraged, whereas all other forms <u>should</u> be condemned, forbidden, or punished. There is no room in this conceptualization for negotiation.

When the term *natural* is used as a prescriptive rule, the term suggests an "ideal nature" and implies the presence of a benevolent supernatural entity like Mother Nature or God,[2] according to historian and classicist John Boswell (1980). Thus, what is *natural*, in this ideal or supernatural sense, is not only good but also moral and in keeping with some divine and preordained plan. Of course, this particular use of the term raises the problem of how one knows God's plan, but the point is really that a set of approved beliefs and sexual behaviors are set apart and sanctified while other attitudes and behaviors are condemned as immoral—as are those who practice these beliefs and behaviors. *Natural*, in the sense of moral and consistent with Mother Nature's or God's plan, often underlies heated discussions about the correctness of certain sexual activities, such as same-sex eroticism, masturbation, and even oral sex in the past. Historically, Christian moralists have asserted that God's plan is that sex is meant for reproduction (Boswell, 1980). Even non-Christians who subscribe to the concept of "ideal nature" often cherish the virtue of reproductive fertility. As a result, both Christians and pagan naturalists generally view vaginal intercourse as *natural* because the activity has the potential for conception. However, for both groups, use of a condom during intercourse interferes with conception and would, therefore, be sterile or *unnatural*. Masturbation and oral sex are also sterile sexual behaviors and *unnatural* in the reproductive sense. By extension of this supposition, it is *natural* for a man and a woman to be together sexually because of their capacity for procreative behavior. Same-sex partners cannot conceive a child no matter what sexual acts they choose to perform together and, so, the coupling of two men or two women is always *unnatural*, as evaluated by reproductive fertility. Yet, by this strict definition of *natural*, <u>any</u> sexual act that does not or cannot lead to conception—even between a man and a woman—would be *unnatural*, including erotic kissing, mutual masturbation, body rubbing, and most instances of vaginal intercourse. Although the Roman Catholic church has generally discouraged and condemned anti-reproductive sexual acts such as masturbation, oral sex, and use of contraceptives, I know of no instances in which it admonished or prosecuted heterosexual couples for (repeatedly) engaging in sexual intercourse that failed to produce children, despite the fact that these sexual acts are sterile.

The term *natural* can also refer to the physical or biological world, rather than an ideal one. In a "real" sense then, what is *natural* is what exists in the world, and what is *unnatural* is what cannot physically exist (Boswell,

1980). In other words, a sexual behavior is *consistent with nature* if it can be found in nature, and the only sexual behavior that is *contrary to nature* in the "real" sense is that which is physically impossible. Thus, for "real" naturalists, heterosexuality, homosexuality, and bisexuality are all *natural* sexual attractions and erotic behaviors because they exist in the world—in the animal world and in human experience. Same-sex eroticism is no less*natural* in the "real" sense than is other-sex erotic attraction, because both exist. Still, as Boswell (1980) points out, those who define *natural* in physical terms often imagine that nature has certain values and is good; consequently, some sexual behavior is bad. A variation of this belief is that what exists without human interference is superior to human behavior or invention. Man-made or artificial materials such as polyurethane, for example, are *unnatural* and bad. Such "real" naturalists devalue sexual behavior that is unique to humans. However, neither other- nor same-sex erotic attraction is artificial or *unnatural* in this "real" sense, because both are evident among animals and humans.

A third association with the term *natural* is the implication that these characteristics are innate, fixed, unlearned, and immutable. When the news media, gay activists, and some scientists note the naturalness of same-sex erotic attraction by referring to "the gay gene" or by stating that "people are born gay," they use the term *natural* in this narrow rigid biological sense. Goodness is then attributed to inborn unlearned traits. Value is attributed to being unalterably created a particular way. Acquired behavior, on the other hand, is disparaged as *unnatural* in that it is somehow less stable, since learned behaviors can be modified or suppressed. The implication is that acquired behavior is weak, fragile, and less good. Many African Americans who oppose gay rights often employ a *natural* argument in this fixed sense (e.g., Williams, 1993). Sociologist and African American John Butler (1993) uses a *natural* argument in the fixed sense when he states that "gay" men and "lesbians" should rightfully be excluded from the United States military and should not be viewed as an oppressed minority who have been denied civil rights. From Butler's point of view, "gay" people are mostly white and affluent and have learned or chosen to engage in same-sex erotic behavior. Butler argues that blacks merit civil rights because skin color is something that one is born with—it is unalterable—but, (white) "gays" do not deserve equal civil rights because homosexuality is a chosen behavior; as an acquired behavior, homosexuality can change; skin color cannot.[3] Butler's assumption is that learned or acquired sexual attraction is somehow under volitional control. Political conservatives and religious fundamentalists of various stripes who accept *a priori* that same-sex erotic attraction is an acquired trait tend also to voice this fixed view of what is *natural*. Good examples of this fixed definition of *natural* by political conservatives can be found in United States Congressman William Dannemeyer's 1989 conspiratorial book, *Shadow in the Land: Homosexuality in America*, and in most public statements about homosexuality made by Senator Jesse Helms (R-NC).

Similar examples of *natural* rhetoric by Christian fundamentalists can be found in James Dobson and Gary Bauer's 1990 book, *Children at Risk*. Dobson, a psychologist by training, heads the multimedia conglomerate Focus on the Family and hosts a national Christian radio program of the same name. Bauer was an advisor to former President Ronald Reagan and now heads the Family Research Council, which sponsored the controversial antigay video series *The Gay Agenda*. Bauer also made unsuccessful attempts to win the Republican party's nomination for president in 1996 and 2000 on an ultra-conservative, Christian Right platform that opposed abortion, gay rights and same-sex marriage, and the so-called "marriage penalty" tax. Obviously, the line between politics and religion is not clearly drawn.[4] Such groups and other like-minded individuals, however, accept without question that other-sex (male-female) erotic attraction is inborn, immutable, and *natural*.

Of course, inherited traits are not always good or fixed. Sickle-cell anemia is an inherited characteristic that is definitely not good. Hodgkin's disease is also not good. Male color blindness and pattern baldness are inherited, and neither of them is a particularly valued trait. Skin color is an inherited trait that is either valued or stigmatized, depending on the particular tonality of skin color and on the social climate. Height is also inherited; again, the value of it—except, perhaps, for extreme tallness or shortness—depends upon the culture. The point is that goodness is *not* relative to genetic inheritance. Inherited traits that contribute to early death or sterility are no advantage, and goodness generally is a socially driven and relative value. Furthermore, inherited traits are not always immutable. Many physical problems or illnesses can be corrected or at least managed with surgery or medication or prosthetics. Diet and medication can control familial hypercholesterolemia, for example. Some heart defects can be corrected surgically, and many people wear glasses to compensate for nearsightedness. The emerging science of genetic medicine also offers new possibilities for manipulating genes to enhance disease resistance and prevent severe health problems.

Furthermore, many acquired characteristics are quite important. Social and parenting skills, persuasion, hunting, or map reading are rare and valued abilities; many people are socially inept, poor parents, and poor providers for their families. Social and communication skills are no less important than inherited traits because they are learned. In fact, the ability to learn and to adapt to a changing environment is the major reason for human survival and dominance. People who believe that homosexuality is learned and therefore bad would probably not otherwise discredit the significance and utility of learning. The Christian values of faith, hope, and love are not inherited characteristics, and these traits are not of lesser value because they are acquired. It should be clear from this discussion that value depends on *what* behavior is in question and *who* is doing the valuing.

The word *normal* is often conflated with *natural* when the topic is sexuality. As mentioned earlier, the two terms are used frequently as synonyms (Schüklenk & Ristow, 1996). *Normal*, however, may refer to a statistical average. The term also means what is common or expected. For example, 5 foot 10 inches may be the *normal* height of American men, although many men are not 5 foot 10. Colds are *normal* in winter months, and rain is *normal* for summer afternoons in New Orleans (where I live). Prior to the present era, same-sex erotic activity was *normal* in this common sense for men under certain social circumstances in most cultures (Bleys, 1995; Boswell, 1980; Cantarella, 1992; Greenberg, 1988; Herdt, 1984/1993a; W. Williams, 1986). So far, none of these references to *normal* imply morality. However, when *normal* is used *prescriptively*, rather than descriptively, the term refers to that which is consistent with a standard— what *ought* to be. Social and religious codes of behavior are usually presented as prescriptive rules to increase conformity. Thus, to say prescriptively that "heterosexuality is natural and normal" communicates the idea that other-sex erotic activity is expected, typical, good, and approved (by society or by God).

Abnormal is the opposite of what is expected or common. *Abnormal* can mean rare, although what is rare could also be highly valued, like virtue or a good deed. *Abnormal* refers more typically to the violation of a (moral) behavioral standard. That is, the paired concepts, *normal-abnormal*, prescribe what ought and ought not to be. In a strictly statistical sense, *normal* sexual behavior for most men in history has been sexual activity with both sexes. However, from a modern prescriptive moral standard of sexual behavior, same-sex erotic behavior is *abnormal* and men who violate this standard are deviants, degenerates, and even criminals. Although being "gay" or "lesbian" is not a crime *per se*, private same-sex erotic activity between consenting adults is illegal in 18 states in America and five of those states define sodomy specifically as same-sex erotic behavior (ACLU, 1999). Historically, "sodomy" has broadly included anal sex, fellatio, mutual masturbation, vaginal or anal penetration with a dildo, *intercrural* (between the thighs) intercourse, and bestiality.[5] During the past 100 years, deviant or *abnormal* sexual attraction has been explained as sickness and psychopathology, and people who engaged in same-sex erotic activity were viewed as mentally ill (Foucault, 1978/1990). Thus, the term *abnormal*, in the moral prescriptive sense, sets apart and stigmatizes people who violate community standards of acceptable sexual behavior.

This long explication of *natural* and *normal* is intended to illustrate that these terms have multiple meanings and are often employed for the moral weight they lend to an argument. They are used prescriptively to convey which sexual behavior is good and what ought to be, according to community standards, which many of us may advocate without even realizing it. However, because *natural* and *normal* can mean many different things, some definitions conflict; one definition condemns same-sex erotic behavior, and another defines it as unremarkable. When these terms are used loosely, same-sex erotic behavior can be all of these things at once—*normal* and *natural* and *abnormal* and *unnatural*. But, pointing out that *natural* and *normal* have multiple and logically inconsistent meanings will not persuade those who believe that same-sex erotic activity is always wrong and immoral. Use of terms like *natural* and *normal* almost never represent a logical argument, but rather a moral one.   Moral prescriptions—as opposed to ethics—do not belong in science and only serve to muddy the issues. As a rule, beware when people talk about what is *natural* and *normal*, because they are probably not describing behavior but telling you instead what you should believe is morally right and wrong.

Throughout this book, I avoid terms that imply goodness or badness or what sexual behavior ought to be. I will not use terms like natural or normal.   My objective is to understand and describe how human sexual attraction to one or both sexes develops. Neither proximate nor ultimate explanations are intended to suggest what sexual behavior ought to be. Hypotheses about the evolutionary function of same-sex erotic attraction are not meant to confer any prescriptive value. Discussions about erotic capacity should not be taken to mean that full capacity must be met or that limits and obstacles to sexual expression are wrong and must be removed. Limits to erotic capacity and unexpressed capacity are functions of a multitude of complex individual, social, environmental, and biological forces; the meaning of erotic limits and unexpressed erotic capacity is left to political activists and religious moralists.   This book is not intended to describe the history of negative beliefs about same-sex eroticism, or condone hurtful social practices, or advocate for more permissive societies. This book attempts to describe a process of development.

Lest my advocacy for a scientific theory of the development and function of sexual attraction implies a general agreement among disciplines, let me note a couple of false assumptions about Science. First, scientists are not in agreement about how human sexual attraction develops. Evolutionary biologists may generally agree about how sexual attraction develops; neuroendocrinologists may agree with most biologists; psychologists may agree amongst themselves but not with biologists; and, sociologists may disagree with everyone but a few like-minded psychologists. There is not, in other words, general agreement across academic disciplines about how sexual attraction develops.   There is in fact a great deal of disagreement among disciplines about the relative contribution of neurobiological, prenatal, postnatal, individual, social, and cultural factors to the development of sexual attraction.   One of the objectives of this book is to synthesize data from diverse fields into a logically consistent *interdisciplinary* interactionist theory. This new theory of sexual attraction is biological and social and psychological, because sexual attraction represents an interaction of events. As I noted earlier, interactionism is messy. Several theorists have proposed interactionist theories of sexual attraction, with limited success. Sigmund Freud (1905b/1953) and John Money (1988) each proposed two of the most influential interactionist theories of sexual attraction. However, I would argue that for the most part the "interactive" piece of Freud and Money's theories has been minimized or ignored by followers who have focused instead on the early intrapsychic experience or on prenatal neurohormonal events, respectively. Michael Bailey and colleagues (1994), David Bem (1996), William Byne and B. Parsons (1993), and Nathaniel McConaghy (1993) also proposed interactionist theories, but these are more specific to same-sex erotic attraction. In general, interactionist theories have captured little public or scientific attention. Granted, it is difficult to articulate neatly how multiple factors combine to produce different sexualities, and previous attempts have not been very "catchy" or even easy to grasp. Nevertheless, an interactionist approach is the one that makes sense and what I present here.

Second, neither Science nor scientists are objective. Scientists attempt to be objective, systematic in their observations, and free of bias in the interpretation of study data. However, scientists are not separate from the society in which they live and work, and cultural values creep into the practice of science (Burr, 1995; Gergen, 1985). Cultural values influence the topics that scientists choose to study and which research questions to ask; social values shape the form of the question and frame what conclusions are drawn from the data (Hubbard & Wald, 1993). As I noted earlier, contemporary Western culture contributed significantly to my interest in sexual attraction and, given another culture, I might be writing about something other than sexuality. Cultural values influence what topics are acceptable areas of study, and in the past scientists who studied sexuality have found that their academic department or university has withdrawn support, or funding sources have obstructed research or demanded that investigators make consequential changes in their project (Bullough, 1994). Edward Laumann and colleagues in their 1994 book, *The Social Organization of Sexuality: Sexual Practices in the United States*, describe the many obstacles they encountered while pursuing government funding for a national survey of adult sexual practices in the late 1980s and early 1990s. Some officials were uncomfortable with the wording of questions, including questions about masturbation. However, the ultimate obstacle came in the form of Senator Jesse Helms (R-NC). Helms, fearing another Kinsey-like study that would normalize the prevalence of same-sex erotic activities and support the agenda of gay rights activists, proposed and won approval of an amendment that forbade funding of the national sex survey. Laumann and colleagues eventually obtained private funding for a scaled-back version of their study.

An uncritical acceptance of cultural beliefs about sexuality by scientists also biases research. The history of the study of homosexuality is an excellent example of how *a priori* beliefs about the psychopathology of same-sex eroticism led scientists to "discover" all sorts of corroborating evidence, which was later found to be unsupported or refuted (Bullough,

1994; Lewes, 1988). Some of these findings are discussed in subsequent chapters. It is important to note that part of the business of scientists is supporting Science, which means that scientists benefit from and cooperate in the enhancement of the institution's social authority. However, when the goal of scientists is the maintenance of Science as an institution of social authority, contradictory and ambiguous data appear to weaken that authority and the pursuit of knowledge is lost. Contradictory and ambiguous data can result from unexamined assumptions and poor methodology. Scientists can strengthen their research by asking themselves how cultural values influence their decision making. Greater attention to how political and cultural attitudes influence research design and the interpretation of data will minimize the likelihood that studies of controversial subjects like sexual attraction will simply reflect back negative cultural beliefs.

1.2 NAMING SEXUAL ATTRACTION

As previously stated, the words we use often have multiple meanings. Common words like *sex*, *attraction*, and *gay*, for example, are particularly prone to multiple and imprecise meanings. To reduce confusion and enhance clarity, I begin with a few definitions of terms and then discuss the hazards of labeling sexual behavior and naming someone's experience.

Sexual activity or *sexual behavior* involves intimate interpersonal physical contact, such as kissing, stroking, genital fondling, masturbation, vaginal or anal penetration, and so on; these behaviors are usually accompanied by specific physiological responses like flushing of the skin, vasocongestion in the genitals, lubrication, and erection. When orgasm is part of sexual activity, pleasurable feelings related to tension release and endorphin activity follow. However, orgasm is not a necessary component of sexual activity, and certain circumstances may prevent some individuals from experiencing sexual activity as pleasurable. A host of feelings besides pleasure may accompany and motivate sexual activity, including fear, sexual pressure or "horniness," being in control, dependency, embarrassment, shame, guilt, affection, and romantic love. In the context of sexual attraction, I am more interested in sexual activity that is motivated by affection and romantic love. While I have identified some behaviors as explicitly sexual (kissing, fondling, and so on), naming a physical act as *sexual* actually depends on the social and cultural context. The same behaviors have different meanings in different situations, some of which may not be perceived as sexual at all. The social context determines *what*, and *when*, and for *whom* physical acts are sexual. Genital contact, for example, may not be defined as *sexual* during play among same-age children, a physician's physical examination, rape, or within some religious or social rituals. The actors themselves may also define the behavior differently. Social status discrepancies between individuals who are physically intimate can cloud the issue of whether and when particular behaviors are *sexual*. I will say more about these circumstantial difficulties later. In short, *sexual activity* is not always a good indicator of sexual attraction. *Sexual activity* occurs for many reasons other than sexual attraction and may take place with individuals other than one's preferred partner or with individuals who are not one's preferred *sex* of partner.

I will attempt not to use the term *sex* to refer collectively to sexual activity. Instead, *sex* refers instead to the biological categories *male* and *female* and their respective secondary physical characteristics. *Gender*, on the other hand, will refer to the socially defined categories of *man* and *woman* and their respective socially derived traits, manners, dress, roles, behaviors, and statuses. *Gender identity* is the personal, private experience of identifying as one sex, while *gender role* will refer to the social roles and behaviors associated with being a *man* or a *woman*. The definitions of *gender identity* and *gender role* employed here differ significantly from how these terms are defined by John Money (1988), their creator. Money views *gender identity* as the personal and private experience of *gender role*, like two sides of the same coin; *gender identity/role* is then the persistent experience of self as male, female, or androgynous, as well as everything that the individual does or says to portray oneself as such, including sexual arousal and response. However, tying gender identity, gender role, and sexual arousal so closely together presents several problems, not the least of which is that one who identifies as female <u>must</u> behave in stereotypically "feminine" ways or risk being labeled pathological. In other words, appropriate gender identity demands acceptance of culturally defined gender roles, although variation in behavior is *not* in itself evidence of psychopathological disturbance in identity. For this reason, I separate *gender identity* and *gender role*.

Attraction is a liking or favoring of a person over others, which may include an emotional commitment. Liking is affected by proximity, familiarity, similarity, reciprocity, complementarity, physical attractiveness, and individual factors such as social status, social skill, and need for affiliation and intimacy. We experience a type of *attraction* and varying degrees of emotional commitment to acquaintances, friends, best friends, lovers, companions, siblings, relatives, offspring, and caregivers such as parents. *Sexual attraction*, then, is a special type of general attraction.

The terms *sexual attraction* and *erotic attraction* are used as synonyms, although the type of attraction that I envision is more erotic than sexual. *Sexual attraction* in my mind is more physically anchored; whereas, *erotic* and *eroticism* includes raw sexual desire as well as more cognitive and social features—emotional intimacy, affiliation, affection, and possibly romantic feelings. However, because the term *erotic* is less familiar and slightly more awkward than *sexual*, I will use the phrase *sexual attraction* to include the erotic. The only exception to this rule is a stylistic one, where the phrase *same-sex erotic attraction* is preferred over *same-sex sexual attraction*.

Sexual attraction refers to an intense intrusive romantic or affectionate longing for physical, emotional and, ultimately, sexual intimacy with a particular individual or members of a sex. *Romantic attraction* is closely related to sexual attraction and refers to the desire for emotional and physical closeness with a favored individual, which may not include sexual activity. Thus, a romantic relationship may not be a sexual one, although romantic attraction denotes a clear (erotic) affection for a particular individual.

Sexual attraction may be felt toward one sex only or to both sexes, although not generally to all members of a sex. Physiological arousal such as a quickening of the heartbeat, nervousness, sweating, or sexual excitement often accompany sexual attraction, particularly when thinking about or in proximity to the favored individual. *Sexual attraction* is first a private emotional experience. However, strong erotic feelings for an individual may be evident in private confessions and even public behavior. Love poems, personal correspondence, literature, plays, personal narratives, law and court testimony, social customs, religious ceremony, philosophical musings, astrological and medical writings, and ethnographic observations are filled with references to sexual and romantic attraction that may not have culminated in sexual behavior. Tapping into multiple informational sources enhances our database and provides a broader understanding of different cultural representations of sexual attraction. Multiple sources of information also serve as a check against misinterpretation of an individual's experience.

To say that someone is sexually attractive suggests that *sexual attraction* is involved. However, *sexual attractiveness* and the phenomenon of *sexual attraction* are not the same, and these terms are not used interchangeably. *Sexual attractiveness* typically refers to an individual's quality of handsomeness or beauty and desirability. More precisely, *sexual attractiveness* is the degree to which an individual or class of persons are viewed as desirable sexual partners. A number of characteristics determine an individual's sexual attractiveness. Evolutionists assert that ancient male-female mating strategies favored several sex-specific physical and behavioral characteristics associated with sexual attractiveness (Geary, 1998). Characteristics such as female youth, large hip-to-waist ratio in females and small hip-to-waist ratio in males, male height and broad shoulders, male social status and material assets, and facial and body symmetry generally advertise good health, lack of genetic defects, female fertility, and male virility and ability to provide physical protection to offspring. Other qualities of *sexual attractiveness*, however, are cultural. For example, what is considered beautiful or virile varies from culture-to-culture and over time. A quick trip through the historical collection of a good art museum is sufficient to identify any number of cultural variations on *sexual attractiveness*. Exemplars of female beauty include but are not restricted to the rotund *Venus of Willendorf* (circa 28,000-25,000 B.C.E.), the rationally proportioned *Aphrodite of Melos* (*Venus de Milo*; circa 150-100 B.C.E.), the ideal porcelain flatness of Edo courtesans in Japanese erotic paintings, and the plumply voluptuous *Hélène Fourment in a Fur Robe* by Peter Paul Rubens (circa 1631), all of whom exhibit a basic female fecundity. Depending on the culture, sexually attractive women are large or small, tall or short, slender or shapely, dark or light skinned, short-haired or long-haired, painted and bejeweled or unadorned, shy or gregarious, or vocal or silent. In recent years, the immense popularity and pervasiveness of Western television, film, fashion, and music has promulgated Western standards of female beauty—white, thin, blond, buxom, and devoid of body hair—which has challenged and in some cases over shadowed native definitions of beauty. Even so, at present, one of the most visible and popular images of female beauty in Western culture is the ultra thin, dark-skinned fashion supermodel, Naomi Campbell.

Cultural definitions of male *sexual attractiveness* also vary, although males are far less often presented as objects of physical beauty. Examples of Western male "beauty" range from Donatello's languidly boyish *David* (circa 1430) to Michelangelo's athletic *David* (1501-1504) and to the contemporary sensuously erotic, buffed, and hairless Calvin Klein underwear models. In most cases, however, masculine attractiveness takes the form of public displays of wealth and bravado.

Besides evolutionary and culture characteristics of *sexual attractiveness*, personal preferences also play a role in the evaluation of sexual attractiveness. Within the larger cultural ideals of attractiveness, some individuals prefer particular characteristics over others—blondes over brunettes, hard muscles over soft flesh, dark colored skin over light skin, gregariousness over coyness, emotionality over even-temperedness, and so on. Thus, *sexual attractiveness* is the degree to which an individual or class of persons possess physical, behavioral, emotional, or personality *characteristics* that are associated with a desirable sexual partner.

For many people, the sex of one's partner is not a variable characteristic—that is, only members of one sex are considered or experienced as potential sexual partners, despite acknowledging the beauty or virility of individuals of the other sex. *Sexual attraction* is not then a function of encountering sexually attractive people. *Sexual attraction* is the degree to which an individual finds one *sex* over the other, or both *sexes*, erotically desirable. Thus, *sexual attractiveness* is a sub-component of sexual attraction, and not the other way around.

Sexual desire also differs from sexual attraction. *Sexual desire* refers to physiological readiness and sexual arousal, (eager) anticipation, and motivation for sexual activity. Sexual motivation is directly related to levels of circulating sex hormones and physiological responsiveness to stimuli. However, whom one finds sexually attractive does not appear to be related to circulating hormones. *Sexual desire* is a general motivational condition, whereas *sexual attraction* directs desire toward one sex but not the other, or to both sexes.

On the other hand, *sexual/erotic orientation* and *sexual attraction* are similar. *Sexual orientation* is a contemporary model of sexual attraction and refers to the experience or capacity for sexual attraction to one or both sexes; this model presumes that direction of attraction is a prominent feature of personality and that various attractions are associated with particular traits. In other words, this model assumes that people *have a sexual orientation* and can be classified by whether they are attracted only to men, only to women, or to both; what is more, each orientation is associated with gender-specific characteristics. Often, *sexual orientation* is described in casual discourse as a dichotomous trait, in which an individual is sexually attracted only to the same-sex or attracted only to the other-sex. *Sexual orientation* has also been described in the scientific literature as a continuum between two poles that represent exclusive same-sex erotic attraction and exclusive other-sex attraction; degrees of attraction to both sexes make up the middle ground between the poles. The continuum model of sexual orientation is consistent with Alfred Kinsey's ratio model of homosexual-heterosexual erotic behavior (Kinsey, Pomeroy, & Martin, 1948; Kinsey, Pomeroy, Martin, & Gebhard, 1953). Although Kinsey initially classified people by the ratio of orgasms experienced with each sex, he later included private sexual fantasy as a dimension of sexual orientation. Other theorists further separated *sexual orientation* from sexual behavior by considering the preferred sex of social affiliates and close friends (Klein, Sepekoff, & Wolf, 1985).

Both the *dichotomous* and *continuum/multi-dimensional* models view *sexual orientation* as an erotic preference for one sex over the other, or for both sexes, that is separate from sexual behavior. That is, a female can have an other-sex erotic orientation as evidenced by exclusive male erotic fantasies and romantic attraction, in the absence of sexual behavior. However, depending on the stage of life or life circumstances, considerable variance and discrepancy can exist across emotional, cognitive, romantic, and behavioral domains in the sex or sexes to which one is attracted, making it difficult to assign individuals to a single discrete category of sexual orientation. For example, it is difficult to classify a self-described "straight" man who has recurrent erotic fantasies about men but who only engages in sexual activity with women and is currently in a romantic relationship with a woman? Is he really "heterosexual"? Is he actually "bisexual" or secretly "gay"? Classifying this guy by a specific sexual orientation depends upon what component of sexual attraction is emphasized. Let us imagine that this same man, after many years of faithful marriage, decides that he is "gay" and begins a romantic relationship with a man; he feels a passionate attraction to men that he does not feel for women, although he deeply loved his female partner. Is this guy now "gay"? Was he "gay" all along? Did his sexual orientation change or just his behavior? And what if we knew that this "gay" guy occasionally fantasizes about sexual relations with women, although he does not pursue sexual relationships with women? How would this information fit into the mix? It is also difficult to classify a woman who identifies as "lesbian" and is in a committed relationship but who on occasion has sexual relations with "gay" or "bisexual" men. Is her sexual orientation really "lesbian," if she has sexual relationships with men? If she only fantasized about sex with men but never acted on it, would her sexual orientation still be "lesbian"?

The conventional dichotomous model of sexual orientation can only explain such variation as a complete change in sexual orientation, or as an earlier suppression or masking of one's "true" sexual orientation. While some versions of the continuum/multi-dimensional model of sexual orientation view such variation in erotic feelings and fantasizes as matters of degree and permit change in sexual behavior over a lifetime, there is much disagreement about how or why these variations occur. For scientific, as well as political reasons, sex theorists, gay activists, bisexual groups, social conservatives, and fundamentalists of various religious beliefs are embroiled in heated exchanges about the flexibility or immutability of sexual orientation. This text offers an explanation for variation and rigidity in sexual orientation.

Sexual orientation is sometimes carelessly used to mean *sexual identity*. However, *sexual orientation* and a chosen *sexual (orientation) identity* label are not the same and not always consistent. *Sexual identity* is the personal identification with one of the conventional categories of sexual orientation: *Heterosexual/straight* or *homosexual/gay or lesbian*. Occasionally, a third category of sexual orientation—bisexual—is used, although the veracity of this label is debatable. An individual may adopt a particular *sexual (orientation) identity* label for political or social reasons, without knowledge of or in full disregard to the inaccuracy of their self-description. Political "lesbians" are such a case; political "lesbians" adopt this label in order to identify closely with women and with issues of concern to women and sometimes to oppose the patriarchal view of women as sexual objects for men. In this sense, "lesbian" is not intended to identify same-sex erotic attraction.

Heterosexual and its colloquial alternative, *straight*, refer to other-sex (male-female) erotic attraction and to people who identify with have this type of attraction. *Bisexual* means sexual attraction to males and females but does not suggest equal attraction to both sexes. *Homosexual* refers to same-sex erotic attraction; however, as a noun, the term carries clinical and negative social implications and has fallen out of favor as a descriptor for a kind of person. For this reason, I will try not to use the term "homosexual," except where meaning would be confused, and will prefer instead the descriptive phrase *same-sex erotic attraction*. At times, I use "homosexual" as an adjective to describe such same-sex attraction but not to describe people. The exception is when use of the term "homosexual" is consistent with a study's sample description or to a given text. For many years now, people who experience same-sex erotic attraction have preferred to identify as "gay" or "lesbian" rather than "homosexual," and this text follows that convention. The terms "gay" and "lesbian" do not carry the same negative baggage as does "homosexual." However, all of these terms refer to a modern conceptualization of sexuality that should not be applied to individuals prior to the late nineteenth century.

Again, *sexual identity* is a personal and political choice and, therefore, *sexual orientation* and *identity* do not always match. Many married men who have sexual relations with other men identify as "heterosexual." Although this group of men could be classified as "bisexual" or even "gay," they seem to be qualitatively different than the small group of men who identify as "bisexual." To separate *sexual identity* from *sexual orientation* in this text, I will use behavioral descriptors such as *same-sex erotic attraction, other-sex erotic activity,* or *sexual attraction to both sexes* when referring to various kinds of sexual attraction.

To emphasize the distinction between identity labels and the study of sexual attraction development, I place *sexual categories in quotation marks*. Encapsulating sexual categories in quotation marks also serves to visually illustrate that these are contemporary social constructs and chosen self-identities, not universal kinds of people. Because our present culture uncritically accepts that "heterosexuals" and "homosexuals" exist as natural kinds of people (and, therefore, have always existed), I wish to remind the reader at every turn that this belief is not fact. What is more, this text is devoted to examining the origin, function, and development of varied sexual attractions, which is very different from a history of the "homosexual" or "heterosexual." Encountering numerous quotation marks throughout the text is awkward and may slow the reader, which fits my purpose. Resisting old and familiar thinking is, unfortunately, cumbersome and uncomfortable, but that kind of resistance is what I hope to facilitate.

Sexuality collectively encompasses sexual health, functioning, desire, attraction, sexual role, sometimes fertility, and usually sexual identity. Since the birth of *sexuality* as a concept in the late 19th century, contemporary Western culture has accepted it as a core component of personal identity (Foucault, 1978/1990; Katz, 1995). Consequently, being a happy, healthy person in Western culture means exercising one's sexuality—identifying with a particular sexual orientation and having an active, satisfying sexual life.

Finally, the term *culture* requires definition. Social experiences and environment are not *culture per se*, which is more than the social interaction between individuals. A friendship, or sexual relationship, or a social role is not a *culture*, although a community of interacting individuals is a culture. *Culture* is the enduring human product of the entwinement and evolution of various social, intellectual, expressive, medical, technological, religious, economic, and political forces. *Culture* is a set of social customs and beliefs; it is a way of organizing and seeing the world. Cultural beliefs are specific to time, place, social status, and sex, although some beliefs have a long history. *Culture* refers to the institutionalization and privileging of social practices such as marriage; *culture* also refers to the richly embedded meaning associated with customs like marriage—social duty, legitimate sexual expression, procreation, role as husband or wife, adulthood status—which makes resistance to social practices nearly impossible. Legal systems are cultural products. The arts are cultural products; cultural beliefs influence the choice of artistic subject, and the creative product reflects and reinforces the culture—even when rebelling against it. Religion and politics are cultural products that represent a convergence of ideas specific to time and place that influence how people in a given culture experience their lives. Infants are born into the culture that shaped their parents' and grandparents' lives and will in turn define their own experience.

In large cultures, values and practices vary distinctly around general themes. Contemporary Western culture, for example, comprises a broad range of beliefs and values that can be subdivided initially into American culture and Western European culture. Within American culture, geographic region, religion, economic class, race, ethnicity, and sex further differentiate social beliefs and practices. Small subcultures that function as relatively closed systems within American society generally incorporate the larger cultural values. Poor, ethnic ghettoized groups are aware of the larger cultural values of, say, social mobility and democracy, even if those practices do not reflect their current way of life.

Culture is the set of beliefs, roles, and institutions that creates our social reality. *Culture* is inescapable. As an influence on individuals and on their experience of sexual attraction, Science is only beginning to understand the enormous effect of culture.

1.2.1 Hazards of Naming Sexual Attraction

Naming *sexual attraction* entails some risk. First, finding a definition of *sexual attraction* that is valid across cultures is difficult, especially if the definition infers some internal experience. Labeling the sexual attraction of someone who is dead or outside our present culture is loaded with risk, as well as possibly arrogant and elitist. Writers who label the sexual attraction of others run the risk of falsely creating eroticism where none existed. Defining *sexual attraction* also sets boundaries that may exclude some erotic experience and present an artificial view of sexual life. On the other hand, studying only sexual behavior tells us what people do but not what they feel sexually, not how they experience their sexual lives, and not how sexual attraction differs from behavior.

Second, language is culture-bound and value-laden. Words have a rich cultural heritage with complex social and political meanings that are transferred, deliberately or inadvertently, to the phenomenon in question. Words are *not* value-free, and the words we choose matters. As noted earlier, people today who experience same-sex erotic attraction prefer to label themselves "gay" or "lesbian" rather than "homosexual." The term "homosexual" was adopted in the late 19th century by psychiatrists and medical practitioners to identify a type of psychopathology and, consequently, this term implies sickness, aberrance, and immaturity. Not surprisingly, people for whom this term was applied find it objectionable. The term "gay" has an altogether different history and different meanings, without the taint of psychopathology. Older definitions of "gay" mean happy or joyful, which challenges the image of disease conjured up by the term "homosexual." The term "gay" is also more encompassing. Whereas, "homosexual" refers to a kind of sexual behavior, "gay" implies more pervasive characteristics or a lifestyle.

Because words and ideas have a historical and cultural context, applying terms outside of their context wrongly creates the impression that there is a direct linkage to the past and that a contemporary sense of the idea existed in the past, when in fact it did not. In his book *Christianity, Social Tolerance and Homosexuality*, historian John Boswell (1980) incorrectly used the phrase "gay people" when referring to people in ancient Rome and in the Middle Ages who had same-sex erotic interests. This contemporary verbiage carries with it many assumptions about sexuality, identity, and individuality that did not exist and not coalesced before the late 19th and early 20th centuries. Although the word "gay" predates "homosexual" by several hundred years (Boswell, 1980), until relatively recently it was not associated with sexual attraction. Therefore, Roman male citizens could not be "gay" or "homosexual," although many Roman men had sex with men or boys; sexual attraction and behavior were simply not defined in this way. Sexuality for Roman men was related to their sociopolitical status, not the sex of their partners. As long as they performed the active sexual role, appropriate sexual partners could be women, boys, or men who played the passive role (Cantarella, 1992). For contemporary Western men, however, the sex of one's partners is extremely important and self-defining.

Here is another illustration of the problems inherent in applying words and ideas out of context: Around the turn of the 20th century, many American and European anthropologists, trained in the new psychoanalytic theory, readily applied this theory to non-Western peoples (Blackwood, 1993). Beliefs about individualism, personal responsibility, the superiority of the white European race, middle-class Victorian gender roles and values, and anxiety about sex were implicit in Freud's theories. As a consequence, psychoanalytic observers overly sexualized and pathologized people who had cultural values and ways of life that were different than Victorian society.

Not only is it a problem to apply contemporary words and ideas to the past, the reverse is also true; unearthing ancient words and concepts and fully grasping their meaning in their original context is a formidable obstacle to modern writers. Beliefs of other cultures often seem strange and irrational, and contemporary values and ideas intrude on their interpretation. An excellent example of this difficulty in understanding ancient ideas is the study of the ancient Greek tradition of *paiderastia*. To contemporary Westerners who believe that sexual relationships should always be between consenting ("heterosexual") adults and that sexual activity with a juvenile connotes abuse, the practice is alien, sinister, and disgusting. The idea that a culture would encourage adult men to have sex with a pubescent boy—although *paiderastia* was much more than a sexual relationship—is abhorrent and clearly wrong to most contemporary Western observers. Yet many cultures sexualize children and sanction sexual relationships between adults and youths. Only recently has Western culture broken from this tradition, although contemporary Westerner observers often fail to recognize this fact. The point, however, is that ideas and words are historically linked and do not transfer easily or well across cultures.

A third caution in naming sexual attraction is related to *who* does the naming. Words and ideas reflect the values and will of the powerful social groups who wield them and alter the way people think about a phenomenon. Words about sexuality such as *homosexual, perversion, promiscuity,* and *contrary to nature* have restricted, regulated, marginalized, and oppressed a particular group of people who were not able to defend themselves (Foucault, 1978/1990). The term "homosexual" still carries many negative associations from the medicopsychiatric practitioners who adopted and popularized the concept during the late 19th and early 20th centuries. The obverse term, "heterosexual," meant *normal sexuality* (Freud, 1905/1953, 1920/1953). The adoption of this binary concept of sexual orientation dramatically changed social reality and continues to do so. What is more, physician's success in authoritatively identifying perverse sexuality established the central and powerful social role of the profession as gatekeeper of psychopathology and regulator of normal sexuality.

Powerful groups other than psychoanalysts and physicians have improved their social position and defined an issue by naming sexual phenomena. Spanish missionaries and conquistadors during the 16th and 17th centuries, for example, saw enormous advantage in naming the sexual behavior of the non-Europeans they encountered. By declaring the indigenous people of the Caribbean and Central and South America to be "sodomites," they redefined social reality (Bleys, 1995). In the Spanish imperialist version of reality, native Indians were barbarian sinners who deserved slavery, rape, wholesale slaughter, the destruction of their culture, and most importantly, the theft of their gold and natural resources. The strength and superior firepower of the Spanish conquistadors and the moral authority of the Catholic missionaries allowed them to reinterpret the meaning of same-sex erotic activities and other sexual behaviors among native Indians and to profit nicely from it. Invading armies, colonists, despots, and other totalitarian regimes gain a great deal by reinterpreting and renaming sexual experience. By approving certain sexual behaviors and labeling other behaviors as immoral, criminal, and intolerable, totalitarian regimes impose control on the sexuality of its citizens, regulate social discourse on sexuality, and increase police power. In 1930's Germany, some of the first and most "effective" acts by Adolf Hitler and his National Socialist Party were the dismissal of women from public office, public encouragement of reproductive (hetero)sexual behavior between Caucasians who were not of Jewish descent, condemnation of non-reproductive sexual activities including masturbation and same-sex erotic behavior, and criminalization of "homosexuals" through a strengthening of Paragraph 175 in 1935.   Such actions created the appearance of nationalism, morality, support of (Aryan) marriage and family, and a determined fight against "evil."   The Aryan German people were encouraged by the National Socialist Party to think of "homosexuals," Jews, gypsies, and other "undesirable" citizens as subhuman or as diseased persons who had infected the national body and needed to be excised. Naming sexual behavior and people in the manner of the German Nazis made it much easier for ordinary citizens to believe that "homosexuals," Jews, gypsies, and foreigners *should* be deprived of their civil rights, stripped of property, shipped to internment camps, exploited as slave labor, and even murdered. Indeed, naming sexual behavior has significant advantages for those who do the naming.

On a lesser extreme but similarly significant level, modern political parties, elected public officials and political candidates, religious leaders, religious fundamentalists, social conservatives, gay rights activists, feminists, civil rights activists, social liberals, scientists, and social constructionists each gain social power and legitimacy by successfully influencing the naming of sexual behavior and people.   At present, these groups are in a battle to determine what particular names and ideas are heard and will be adopted by society. Will people who are attracted to members of the same-sex be viewed by society as *sinners, perverts, criminals,* or as *an oppressed minority*? Will heterosexuality be viewed as a superior social position and exemplary model for society, or will it be seen as only one of many possible ways to structure society? Which (group's) viewpoint will dominate? For the past 20 years, social constructionism has largely dominated the field of sexuality, which was dominated previously by medical science. Gay activists, feminists, gender theorists, sociologists, and university faculty members with a social constructionist perspective have, for better or worse, directed intellectual discourse on sexuality in recent years. As a result, constructionists have benefited directly through journal and book publications, public acceptance of their ideas, public recognition, faculty appointments, and a strong student following. In no small way, the outcome of naming contemporary sexual behavior has real consequences for each group involved in this struggle— including social constructionists—and for all of us.

Finally, social groups also influence social reality by not *correctly* naming a phenomenon and by deliberately *mis*naming it. This may seem to be a subtle or redundant point, but it is not. For many years, *revisionist* historians, scholars, and advocates of popular conventional culture refused to acknowledge the same-sex erotic experiences of prominent historical figures and whole cultures or refashioned those experiences to reflect contemporary "heterosexual" ideals. Before the middle part of the 20th century, American and European anthropologists largely refused to mention the existence of same-sex erotic activities in non-Western cultures (Blackwood, 1993). For those anthropologists who even recognized same-sex erotic behavior, such behavior often escaped objective description and analysis out of embarrassment, disgust, ignorance, fear of promoting such behavior, and fear of personal moral condemnation and academic censure. Anthropologists themselves were not solely responsible for failing to mention same-sex eroticism; pressure from colleagues, academicians, and publishers had a powerful influence the omission or revision of observations by scholars. The social context of the first half of the 20th century did not permit an open discussion about same-sex eroticism even within academic circles; same-sex erotic behaviors were viewed as perverted, immoral, and contagious. The moralistic Cambridge dean in E. M. Forster's 1913 semiautobiographical novel *Maurice* sharply illustrates this view when he advises a student who translates a portion of Latin text to "omit" a reference to same-sex eroticism, "the unspeakable vice of the Greeks" (1971, p. 51).[6]

One way that historical and cultural references to same-sex erotic attraction have been ignored, omitted, and mislabeled by scholars is by setting impossibly high standards for concluding its existence (Boswell, 1990; Norton, 1997). No such standards are required to conclude other-sex (male-female) eroticism or sexual activity. Scholars readily and eagerly infer and speculate about other-sex erotic attraction and sexual behavior from slim evidence and casual encounters. Yet, in the absence of "hard" proof (and even at times in its obvious presence) of genital-genital contact or orgasm, same-sex eroticism cannot be determined with absolute confidence and, therefore, is not even mentioned. Without evidence of an ejaculating penis or an orgasming vagina, many scholars have claimed that intimate affections, extraordinary violations of social custom, and long-term primary relationships between same-sex partners in the past and in other cultures were *not* sexual, *not* erotic, and merely close friendships. I am not arguing that same-sex eroticism, (perhaps) a statistically less common event than other-sex eroticism, should be read into every same-sex intimate pairing. I do argue, however, that evidence of same-sex erotic attraction and behavior has been ignored because it fails to meet a narrow definition of *sexuality as genitally focused*. The very idea that genital contact is required as evidence of sexual behavior and that sexual behavior, so defined, is required to identify a passionate, romantic relationship is a contemporary Western fiction. In other cultures, and even in our own, genital contact or orgasm need not be present in erotic relationships, which are often characterized by intimate physical contact, preference, altruism, tenderness, emotional support, affection, devotion, fidelity, and jealousy. In my recent academic training, I have witnessed respected scholars describe the ancient Greek tradition of *paiderastia* as a non-sexual, non-erotic, passionless social custom that was more form than content, *in spite of considerable evidence to the contrary*. While eroticism and sexuality are defined differently by other cultures, these phenomena certainly exist and infuse many relationships. What is more, this scholarly bias against identifying <u>any</u> same-sex erotic attraction in pre-industrial non-Western cultures strikes me as parallel to what pro-gay activists do when they see "gay" sexuality in every same-sex encounter or reference.

The outcome of this bias against naming same-sex eroticism in pre-industrial non-Western cultures is a sanitized, de-(homo)sexualized interpretation of history. In such a context, the erotic lives of individuals and cultures in the past are rewritten to conform to contemporary beliefs and ideals. A good example of revisionist biography can be found in Michelangelo the Younger's 1623 collection of his great-uncle's poems. Prior to publication, the great-nephew changed all masculine pronouns in his great uncle's poems to feminine pronouns, creating the false impression that Michelangelo loved women rather than men (Norton, 1997). The deception went undiscovered for almost 250 years. However, in 1863 historian John Addington Symonds found a note written in the margins of the poems by Michaelangelo's great-nephew that stated that the poems should not be published in their original form because they expressed "*amor...virile*."

Another outrageous example of the heterosexualization of historical figures is the fabrication of Sappho's marriage to Cercylas (Gettone, 1990). What is more, Ovid portrayed Sappho as the lover of the sailor Phaon, a completely fictional character. This fiction of Sappho's romantic relationships with men was repeated and reinforced for centuries. In another infamous case, Vasco da Lucena resexed the eunuch Bagoas, Alexander's male favorite, as a beautiful young woman in his medieval French translation of ancient Roman Quintus Curtius Rufus's *History of Alexander the Great* (Norton, 1997). During that period in France, depicting a great man like Alexander—a model for all men—in an erotic relationship with another man would have been unthinkable; apparently, da Lucena thought that a lie was more acceptable than the truth.

In recent examples of revisionist history, Queen Christina of Sweden, played by woman-loving Greta Garbo, was heterosexualized in the popular 1933 Hollywood film of the same name (Russo, 1981). In the film, despite a fictionalized love affair with the Spanish ambassador, Queen Christina refuses to marry and abdicates her throne. In reality, however, she refused to marry *any* man and was forced to abdicate. Records suggest that Christina loved the countess Ebba Sparre. Similarly, in the 1962 film *Lawrence of Arabia*, Lawrence, played by Peter O'Toole, withstands a fictionalized night of torture at the hands of the evil Turkish bey (Jose Ferrer), although Lawrence himself implied that his experience with the Turkish ruler was sexual and pleasurable rather than abusive (Russo, 1981). The film makes no mention of Lawrence's sexual attraction to or erotic relationships with men. As more recent examples of contemporary heterosexualization, movie studios have often staged public "dates" with beautiful women for their male stars when speculation arose about their homosexuality.   In the 1950s, a major movie studio forced Rock Hudson to marry Phyllis Gates in an attempt to quash rumors that he was gay.   The studio feared a huge loss of revenue if their sexy masculine star was known to prefer men as sexual partners.   While the studio was probably more concerned with their own profits than Hudson's same-sex erotic behavior, the social effect of renaming his sexuality was the same.   In 1985, stricken with AIDS and near death, Hudson reluctantly acknowledged that he was "gay."

In these few examples, same-sex eroticism is denied and redefined as other-sex eroticism.   By not acknowledging and not naming same-sex eroticism, scholars and social groups cover over evidence of same-sex erotic behavior.   By renaming same-sex eroticism as other-sex eroticism, scholars and social groups fabricate history, restrict public discussion about same-sex eroticism and drive it underground, and make the public promotion of conventional male-female sexuality a political agenda.

"Gay" men and "lesbians" who portray themselves as "heterosexual" are not revisionists in this sense.   Although for both "gay" people and revisionists fear and discomfort with same-sex eroticism underlie such deceptions, the motives of revisionist historians, profit-hungry movie executives, and individuals with same-sex erotic feelings are not the same as for "gays."   "Gay" people misrepresent themselves as "heterosexuals" in order to avoid discrimination, ridicule, or violence; revisionists, on the other hand, rewrite the sexual lives of others because there are offended by same-sex eroticism, want not to offend others, or wish to set a good example for society.

Too seldom, scientists have failed to consider the implications of naming or not naming erotic experience.   The interpretation and presentation of sexuality has political consequences for informants, observers, and historians and should be undertaken with great caution.   The words we use create social reality and often reinforce old social prejudices.   My intention is not to construct a new social reality in this text but to describe erotic experience.   For that reason, I prefer using descriptive language and will refer to sexual attractions and behavior rather than to labels such as "gay," "lesbian," "homosexual," "bisexual," or "heterosexual," unless I am specifically referring to people who identify with one of those labels.   I also recognize that outside observers of a culture and cultural critics are generally disposed to viewing erotic behavior in ways that match their own social values.   As a result, many reports about sexual behavior in non-Western cultures are biased toward Western values.   Therefore, conclusions drawn from non-Western cultures must be tempered.   Greater awareness of my own cultural biases and the biases of other writers may help me avoid common missteps and facilitate a thoughtful description of the development of human sexual attraction.

1.3 OVERVIEW OF THE BOOK

Linguistic quicksand is pervasive in the field of sexuality. Words have multiple meanings, and the choice of words has political consequences. I will not use words lightly. And, I proceed with caution, knowing full well that an interactionist theory of human sexual attraction is unattractive and theoretically muddy. Explaining sexual attraction in terms of probable causes and multiple cofactors makes for a less colorful story than pronouncing absolute certainty of a clear single cause. Interactionist theories of sexual attraction are difficult to express in easily grasped phrases or catchy sound bites. However, when grappling with understanding complex human phenomena, absolute clarity is unlikely. We as scientists are certain about very little regarding human sexuality, but uncertainty is exactly the reason for further investigation and theorizing.

Current theories of sexuality are stagnant and uninspiring. It is time to reach beyond the safety of what we know for sure and speculate about what may be. This intellectual stretch may prove futile or may inspire a new vein of creative thought and research. In my opinion, an interactionist theory of sexual attraction provides the fertile ground necessary for the growth of new ideas about sexuality.

To state my thesis briefly, I hypothesize that during human evolution sexual desire became socialized and that varied sexual attractions benefited individual survival, and ultimately, reproductive success. As our early human ancestors developed complex social environments, it is likely that sexual desire acquired social in addition to reproductive functions, which were no less vital to survival. The increasing social complexities of early hominid collectives socialized sex drive. In an environment organized by social relationships, the desire for intimate (sexual) companionship facilitated the development of sexual attraction from sex drive. As sex drive was overlaid with layers of social meaning and cognitive expectations, it became more conscious, varied, and socially structured. Sexual attraction then became a more prominent feature of social life, even though most instances of it did not result in sexual contact. I suggest that eroticism initiated, fueled, and maintained successful friendships, as well as sexual relationships. The demands for skillfully building cooperative social and sexual relationships and managing multiple social roles are likely to have promoted variance in whatever "glue" held these alliances together. In other words, well-played eroticism to *both* sexes gave those individuals a social advantage and contributed to their survival. Culture and social context defined which events were considered sexual, appropriate, and allowed or taboo. What is more, culture defined emotional (including erotic) experience, setting limits on what was possible and appropriate to feel and how feelings should be expressed. Thus, social life created boundaries for sexual experience. In short, sexual experience is a product of behavioral opportunity, the individual's history and personal preferences, and his or her cultural context. The individual's constitution and the cultural context interact—or interface—at the social, emotional, cognitive, and neurochemical levels. All stimuli are processed, stored, and recalled as neurochemical messages; therefore, human sexual attraction is an interaction between external events and individual internal neurochemical processes and experience.

Data regarding neurohormonal processes and structural and functional brain differences between individuals of different sexual orientations suggest a possible sequence of events in the development of sexual attraction. Variation in the direction and intensity of sexual attraction is likely to be a function of the variable sensitivity of brain cells to particular hormones and the timing, quantity, and duration of an individual's exposure to prenatal and postnatal sex hormones. Differential patterns of pre- and postnatal hormone exposure create variability in the processing of social stimuli. Over time, differential processing of erotic stimuli solidifies dispositions or traits, or informational processing is obstructed, depending on the social and sexual opportunities presented by the culture and on what traits and forms of attraction are encouraged. A given culture fosters some traits and leaves others undeveloped; social practices and institutions support approved kinds of sexual behavior and condemn socially unacceptable behaviors.

Thus, human sexual attraction can be thought of as the product of genes *and* sex hormones *and* social learning *and* active participation in a socially constructed world. Isolating the individual effects of genes or of social learning is impossible at present. Deciding where hormonal influences end and social experiences begin is guesswork, given our current technology. However, I argue that identification of the precise individual contributions to sexual attraction is not required for an understanding of the interactive model.

Now that I have outlined my thesis, let me say a bit about the layout of this book. Chapters 2, 3, and 4 describe and critique conventional theories of sexual attraction. Many of these are theories of male same-sex erotic attraction. Where possible I extrapolate these theories to other forms of sexual attraction and include more general theories of attraction. Most conventional theories of sexual attraction are linear and involve single or relatively few causal factors. Chapter 2 presents two important psychosocial models of sexuality—psychoanalytic and learning models. In spite of different premises, psychoanalytic and learning models share many common features. Chapter 3 discusses an extreme social constructionist model of sexual attraction through a one-sided "Debate" by radical constructionists with their absent opponents—radical biomedical theorists. This quirky presentation was chosen because this is the approach most often employed by social constructionists themselves for presenting their views on sexual attraction. Social constructionism is a complex amalgam of related ideas that contrast sharply with so-called essentialist biomedical theories. The model offers provocative explanations of the development of sexual categorization that find some support in the social evolution of Western sexuality. Chapter 4 describes several biological theories and numerous research studies on sexual attraction. Unfortunately for the reader, biological research on sexual attraction spans a wide technical gulf. Early sex research studies largely developed generalizations from the physical development and sexual behavior of nonhumans. More recent studies, however, have employed sophisticated medical technology to examine human brain structures, sex hormone levels, and cognitive functioning.

Support for each major theory (or the lack of it) is sought in cross-cultural and historical examples of sexual relationships and in different concepts of sexuality. In general, the persistence and resilience of same-sex eroticism—despite diverse social attitudes, politics, and definitions of sexuality—suggests the presence of a biological trait. Following this critique, critical elements from different perspectives are salvaged and reorganized in an interactionist theory of sexual attraction.

Chapter 5 provides a more comprehensive interpretation of current knowledge about sexuality and presents a new interactionist model of general sexual attraction. I describe how the increasing importance of social reality among early human societies challenged the brain and affected sexual life. I suggest that the socialization of sex drive promoted variance in sexual attraction and elevated the significance of sexuality in the social world.

Finally, Chapter 6 deals with present and future questions about sexual attraction. This section outlines an ambitious research program to test the new model of sexual attraction.

It is my hope that the rudimentary hypotheses raised in this book will stimulate more erudite proposals and a new stream of knowledge about sexual attraction.

CHAPTER 2: LIBIDO AND LEARNING: PSYCHOSOCIAL MODELS

Homosexuality is an abomination to the Lord. . . . Homosexuals will recruit our children. . . . [Because gays cannot have children], they can only recruit children, and this is what they want to do. Some of the stories I could tell you of child recruitment and child abuse by homosexuals would turn your stomach. — Anita Bryant

He's all man—we made sure of that. — Ronald Reagan in reference to his son, Ron

Same-sex eroticism is a problem for most moralists and many scientists. As this chapter's epigraphs suggest, same-sex erotic attraction is often seen as a contagion, and people who have such desires are seen sometimes as dangerous predators of innocent children. Even people who believe that same-sex erotic attraction is inherited may worry that children will learn to be "gay" from hearing about "gay" people's lives, watching "gay" characters on television, or having intimate physical contact with someone of the same-sex. Many religious fundamentalists of various stripes—Christian, Islamic, and ultra-orthodox Judaism—believe that same-sex erotic attraction is easily acquired and a deadly threat. Some Christian fundamentalists describe same-sex eroticism as a potent virus[1] that is strategically designed to break through fragile "heterosexual" defenses and infect and destroy "our" most cherished (heterosexual) traditions (Dannemeyer, 1989; Dobson & Bauer, 1990). In fact, fear of same-sex erotic attraction is at the core of contemporary Western culture. For over 100 years, Western culture has obsessed about male same-sex eroticism, which is the primary reason why scientific theories of erotic attraction are largely theories of male homosexuality. Male-male eroticism has been seen in this culture as

31

a critical social and moral problem that demands explanation and remedy. Female-female eroticism, on the other hand, has been far less of a cultural concern and has received little attention, scientific or otherwise. And, male-female eroticism, while requiring defense against deviant forms of sexuality, has been viewed as normal, natural, automatic, and needing no explanation.

Over the past 100 years, the scientific community has echoed and magnified these cultural beliefs about same-sex male eroticism. Despite pronouncements of cultural objectivity, scientists are very much a part of their culture—not apart from it—and their social assumptions often mirror current prejudices (Kuhn, 1962/1996). Harvard professor Ruth Hubbard supports the notion of subjective Science when she notes that "scientists, as a group, tend to provide results that support the basic values of their society. This is not surprising, since scientists live in that society and make observations with that society's eyes. . . . This is particularly true when scientists are studying people (Hubbard & Wald, 1993, p. 7)." Because scientists are representatives of their culture and often hold unexamined cultural beliefs, it should not be surprising that most theories of sexual attraction that we discuss reflect social prejudices about homosexuality.

Scientists, as well as civil authorities and spiritual leaders, have believed at different times that male-male eroticism is inherent or a choice, congenital or learned, sickness or sin, depending on prevailing social attitudes. Interactions of biological or social forces are almost never mentioned as an explanation for same-sex eroticism. Many social scientists in the past assumed *a priori* that same-sex male eroticism is a social problem and, as such, must be either inherited or acquired. Framing the research question in this way accepts negative social beliefs about a group of people and reflects the desire to find a simple causal explanation for the phenomenon and to place people neatly in exclusive categories. However, the pursuit of a *single* cause to confirm the assumption of psychopathology produced confusing, contradictory, and often meaningless data. The lack of methodological rigor among sex researchers worsened the problem. If we exclude all studies with small samples, poor assessment instruments, and researchers who distorted data to validate a malignant view of same-sex male eroticism (as many psychoanalysts did in the two decades following the Second World War), few studies are left that tell us anything useful about male-male sexual attraction. However, a number of colorful and imaginative theories of sexuality remain.

Before reviewing learning theories, let's consider just how big a "problem"—how prevalent—same-sex erotic attraction is. How prevalent are other forms of sexual attraction? Alfred Kinsey (Kinsey, Pomeroy, & Martin, 1948; Kinsey, Pomeroy, Martin, & Gebhard, 1953) reported that 50% of men and 70% of women were exclusively "heterosexual," as defined by their sexual activity. That so few men were exclusively "heterosexual" was shocking to say the least. Only 4% of men and 2% of women were classified as exclusively "homosexual." But these were not Kinsey's most controversial findings. He also reported that 37% of men and 13% of women had had at least one same-sex experience to orgasm as adults. Another 13% of men and 10% of women had same-sex erotic fantasies but did not engage in sexual behavior. Collectively, these findings set off a firestorm of protest that has never completely died down. Without rehashing the methodological problems with Kinsey's studies, most social scientists now conclude that the frequency of same-sex experiences reported by Kinsey was inflated as a result of a biased sample. How then does Kinsey's data square with other studies?

In the 1970s, Shere Hite conducted a large convenience survey, entitled *The Hite Reports* (1976, 1981), and found that 85% of men and 79% of women identified exclusively as "heterosexual." About 11% of men and 8% of women identified as "homosexual," and 4% and 13% of men and women, respectively, called themselves "bisexual." A second large study, *The Janus Report* (Janus & Janus, 1993), based on a national convenience sample and interviews with 2,765 Americans, reported that 91% of men and 95% of women identified as "heterosexual." Four percent of men and 2% of women—a much smaller proportion than in the Hite studies—described themselves as "homosexual." Only 5% and 3% of men and women, respectively, identified as "bisexual." Convenience samples, like *The Hite Reports* and the *Janus Report*, suffer from similar methodological biases as the Kinsey's studies.

More recently, however, Edward Laumann, John Gagnon, Robert Michael, and Stuart Michaels (1994) used probability sampling to obtain data on 3,159 representative Americans. Within this sample, 97% of men and 99% of women identified as "heterosexual." A tiny 1% and 2% of participants called themselves "gay" or "lesbian," respectively. About 4% of men and women reported being sexually attracted to both sexes. Self-identity, however, did not correspond completely to sexual behavior. That is, 4% of men and over 2% of women had enjoyed at least one same-sex encounter to orgasm during the past five years; 10% of men and 4% of women had experienced a same-sex encounter to orgasm in their lifetime. In other words, 10 times as many men and twice as many women had engaged in same-sex activity to orgasm in their lifetime as those who identified as "gay" or "lesbian."

Using a 1970 probability data set of 1,450 adults compiled by the Kinsey Institute, researchers Fay, Turner, Klassen, and Gagnon (1989) found that between 3% and 6% of adult men had *regular* same-sex erotic encounters. An estimated 20% of men had one or more same-sex erotic experiences in their lifetime. Married men made up the largest percentage of men who had same-sex experiences. European probability studies have obtained somewhat lower figures for same-sex behavior. Over 4% of men and nearly 3% of women in France reported a same-sex erotic experience in the past five years (ACSF, 1992). More than 6% of men and nearly 4% of women in Britain had one or more same-sex experiences to orgasm in their lifetime (Johnson, Wadsworth, Wellings, Bradshaw, & Field, 1992).

The probability studies in Europe and America draw similar conclusions from their data (ACSF, 1992; Fay *et al.*, 1989; Johnson *et al.*, 1992; Johnson, Wadsworth, Wellings, & Field, 1994; Laumann, Gagnon, Michael, & Michaels, 1994), most of which are also consistent with large convenience studies (Hite, 1976, 1981; Janus & Janus, 1993; Kinsey *et al.*, 1948; Kinsey *et al.*, 1953). First, most people self-identify as "heterosexual." Second, self-identity does not always correspond with sexual behavior; a number of men and women have some sexual encounters that contradict their self-identity. Third, a significant proportion of men and women report sexual fantasies about or attraction to both sexes. Fourth, an average of about 5% of adult American men and 3% of women regularly engage in same-sex erotic activity. Fifth, men are more likely than women to engage in same-sex erotic activity. Sixth, same-sex erotic behavior cuts across age, education, race, economic status, occupation, religion, and political affiliation. And in general, social stigma associated with homosexuality may contribute to underreporting. Therefore, data about same-sex eroticism may represent minimum proportions.

In all, it seems safe to conclude that same-sex erotic attraction is not an epidemic and does not deserve all the attention given to it over the past 100 years. Across samples, only a few people—mostly men—are exclusively interested in the same sex, although a number of people have a same-sex erotic experience to orgasm at least once in their lifetime. Most people identify as exclusively "heterosexual," even though some acknowledge having same-sex erotic fantasies or engaging in same-sex erotic behavior. Of course, the problem for moralists and scientists is that quite a few people do not stay neatly in their sexual category—that is, they call themselves one thing and do other things as well. This problem is unique to contemporary Western culture. Although many cultures have recognized that few individuals are exclusively interested in one sex, sexual attraction was not defined as a personal identity. The ancient Greeks and Romans, the Japanese prior to the late 19th or early 20th centuries, Melanesian societies in the south Pacific, and early Native American Indians conceive of sexual attraction somewhat differently. However, none of these cultures view the sex to whom one is attracted as an important characteristic of identity. Deciding who is "gay," or "straight," or "bisexual," and why, is largely a contemporary

Western concern.

One set of theories about how sexual attraction develops has to do with psychological experience and social learning. Here I review the psychosocial theories of attraction that have most influential in the past century. These theories are, as I have said, largely theories about male-male sexual attraction. When possible, I extrapolate from a same-sex theory to suggest how other-sex attraction and attraction to both sexes may be explained, or not. The major psychosocial theories of sexual attraction fall into two groups: Freudian and post-Freudian psychoanalytic approaches and behavioral conditioning models. I address each in turn.

2.1 FREUD AND THE POST-FREUDIANS

2.1.1 Freudian Theories

Neurologist and psychoanalyst Sigmund Freud produced several theories of erotic attraction, which are scattered throughout his hefty body of work. Freud described the origin of "normal" sexuality—his word for male-female eroticism. He also proposed one loose theory of same-sex female erotic attraction and four theories of male-male attraction. The different theories of same-sex male eroticism are not so much an evolution of ideas but, rather, an illustration of Freud's struggle to grasp the complexity of the phenomenon (Lewes, 1988). Freud (1910/1953, 1920/1953) readily acknowledged that his inexact theories failed to explain "the various forms of homosexuality" and were founded on a small number of persons.

Most of Freud's thoughts about the development of erotic attraction are found in three primary works: *Three Essays on the Theory of Sexuality* (1905b/1953), "The Psychogenesis of a Case of Homosexuality in a Woman" (1920/1953), and "Certain Neurotic Mechanisms in Jealousy, Paranoia and Homosexuality" (1922/1953). Although he remained relatively ambivalent about whether same-sex erotic attraction was a psychopathology, Freud (1905b/1953) nonetheless described same-sex eroticism as an inhibition of "normal" psychosexual development. He considered male-female erotic attraction to be the primary outcome of healthy psychosexual development. However, Freud distinguished between sexual "perversion," which was pathological but not inclusive of same-sex eroticism, and "inversion" (his frequently used word for homosexuality), which was not pathological (Lewes, 1988). In his oft-cited letter to an American mother who was worried about her son's same-sex attraction, Freud stated in unambiguous terms that "homosexuality is assuredly no advantage, but it is nothing to be ashamed of, no vice, no degradation, it cannot be classified as an illness; we consider it to be a variation of the sexual function produced by a certain arrest of sexual development" (cited in Lewes, 1988, p. 32). All the same, same-sex erotic attraction was still problematic for Freud.

Freud's first theory of male same-sex eroticism was derived from the case of Little Hans (1909/1953, 1910/1953). According to this theory, a boy loves his mother and views father as a rival, fearing castration as punishment from his father. Eventually, the boy discovers that mother has no penis—and responds in horror and shock. He is disgusted and fearful of mother's genitals and severs erotic ties to her. As an adult, the man chooses as his compromise love object a woman with a penis or, rather, a feminine-looking male. Each subsequent sexual encounter with an effeminate male reinforces his belief in the possibility of finding a woman with a penis and recalls his earlier erotic bond with mother. Thus, this theory predicts that men who love men want effeminate men as partners. Loving a masculine man would challenge the implicit desire for a woman with a penis and create anxiety. However, a systematic review of available literature reveals no supportive evidence for the supposition that men who love men desire feminine-looking men. On the other hand, there is considerable literary and anecdotal evidence to the contrary. A few minutes in a gay bar watching the audience's reaction to a buff male stripper should completely dispel this notion.

Freud's second theory of male same-sex eroticism, described in *Three Essays*, is distinct from but consistent with the first. This in fact is the theory most often repeated by Freud (Lewes, 1988). According to this theory, a boy has a lengthy emotional narcissistic relationship with his mother, his first love object. Father is again the rival. In his long non-sexual erotic relationship with mother, the boy comes to overvalue his penis. Consistent with the first theory of male same-sex eroticism, the boy fears castration by father and is disgusted and shocked when he discovers mother's penisless state. The boy avoids castration anxiety and disgust by directing his erotic feelings away from mother. After the traumatic separation from her, the boy preserves this intense relationship by identifying with mother, as the object of affection, but choosing partners to love who resemble himself. By loving males, the adult man re-creates and re-experiences the original erotic bond with mother, as the loved object—in effect, loving the mother in himself and being loved. For this man, male sexual partners are substitutes for the unattainable mother love object and a symbol of himself. Unfortunately, *introjections* are difficult to test because they are unconscious and symbolic.

And, the problem with symbols is that if one wants to see them, they seem to be everywhere.

The third theory of same-sex male erotic attraction was derived from the case of the Wolf Man, although he was not "homosexual" (1918/1953). For Freud, this form of same-sex attraction was an unconscious erotic identification—a sort of negative *Oedipal complex* (Lewes, 1988). In the case of the Wolf Man, the father was the object of masculine identification for the boy. According to Freud's third theory of same-sex erotic attraction, the boy gives up his masculine identity in order to be loved by his father in the same way that his father loves women. However, the boy avoids becoming "homosexual" by abandoning his desire for father's love and keeping his masculinity. Freud viewed this type of unconscious same-sex male eroticism as more common than all other forms of homosexuality. Father figures often recur in a man's life and, when they arise, the struggle between wanting "father's" love and affection and keeping one's (heterosexual) masculinity is replayed. Thus, simply put, Freud's theory suggests that "gay" men want to be loved sexually by their father and that "heterosexual" men do not. This view is consistent with Western male socialization not to express tender feelings and to fear male-male affection. The expression of tender feelings toward another male, even one's own son, may carry the taint of homosexuality for many men who then avoid displays of affection toward their son. It is also probably true that most men want their father's love and feel that they have received too little affection from him. Indeed, some evidence suggests that fathers reject or distance themselves from their "gay" sons (Isay, 1989). However, the research literature provides no support for the contention that "gay" men want their father's love more than do "heterosexual" men or that "heterosexual" men who seek their father's affection are latent "homosexuals."

Freud mentioned his fourth theory of male-male eroticism only once and in relationship to explaining paranoia and jealousy (1922/1953). According to this theory, a boy develops an intense emotional relationship with mother and becomes jealous of her attention to his siblings and father. His violent wishes to harm his rivals for mother's affection are transformed through the process of *reaction formation* and the rivals become love objects. Thus, the violent wish to harm becomes a heavy-handed desire to help. Freud speculated that overly altruistic behavior by men who love men masks violent fantasies. Unlike the other forms of male same-sex erotic attraction previously discussed, this one occurs earlier in development and without fear of female genitalia. This theory is particularly interesting because sociobiologists have speculated that "gay" men (and perhaps "lesbians") are selected by parents not to reproduce in order to lessen competition for resources among kin and enhance the chances of survival for siblings' offspring (Wilson, 1975). In giving up their chances to have children and devoting their resources to the survival of their siblings' children, "gay" people are *reproductively altruistic* (Weinrich, 1987a). There is, of course, ample evidence that "gay" men and "lesbians" have fewer children than "heterosexuals." However, there is not strong evidence to show that "gay" people are any more altruistic than "heterosexuals."

In contrast to his rich ideas about male-male eroticism, Freud offered one account of same-sex female erotic attraction. Through analysis with a "lesbian" patient, Freud (1920/1953) formed his only theory of same-sex female eroticism. He speculated that, in deeply loving her father, a girl wants a child by him and hates her mother. When father does not provide her a child, the girl rejects not only father but also all men and becomes a "man" in the active sexual sense by taking her mother as a love object. As an adult, the woman searches for a mother-substitute in other women. This particular theory was not well developed and not readily accepted by Freud's followers. In contrast to his theories of male-male eroticism, Freud's theory of same-sex female eroticism was relatively ignored. This fact supports the notion that the idea of males violating male social roles—and not same-sex eroticism in general—is what most bothered and enraged phallocentric psychoanalysts and Western society.

Despite minimal interest in female-female eroticism, Freud viewed same-sex eroticism in general as complex and varied. He acknowledged his failure to provide a comprehensive explanation of same-sex erotic attraction but believed, however, that constitutional or biological factors contributed significantly to the development of same-sex eroticism.

Despite, in fact, a great of interest in and theorizing about sexuality, Freud viewed the entire phenomenon of human sexual attraction, including "normal" sexual attraction, as a puzzle begging to be solved. Freud (1905b/1953) wrote at the turn of the 20th century, "In the psychoanalytic sense the exclusive interest of the man for the woman is also a problem requiring an explanation, and is not something that is self-evident" (Lewes, 1988, p. 35). In his conceptualization of psychosexual development, Freud assumed that all children are *polymorphous perverse*, meaning that a child's emotional capacity is *pan*-erotic and capable of a wide range of sexual responsiveness. In his mind, children are not born with a directed sexual attraction; they are sexually attracted to and stimulated by almost everything. This pan-erotic capacity is directed and narrowed through a series of intrapsychic conflicts during childhood, resulting ultimately in other-sex erotic attraction and behavior. Freud's theory of psychosexual development does *not* allow for a healthy adult to be attracted to both sexes or to the same sex.

Freud's classic theory of male (hetero)sexual development was derived from Greek myth, although his theory and the myth emphasize very different points. The Greek tragedy of Sophocles' *Oedipus Rex* describes the folly of trying to avoid fate. In the play by Sophocles, Apollo tells King Laius of Thebes that his son Oedipus will kill him. Laius leaves his infant son on a mountain to die, but the infant is rescued and raised by King Polybus of Corinth. Later, the oracle tells the adult Oedipus that he will kill his father and have intercourse with his mother. Oedipus flees to Thebes to avoid his fate but, on the way, he encounters a rude a carriage driver. In a fit of rage, Oedipus slaughters everyone in the group, including the man in the carriage, his father, King Laius. Unaware that he has already fulfilled half his fate, Oedipus arrives in Thebes to find the city bullied by the Sphinx. He solves the Sphinx's riddle, saves the city, is made king of Thebes, and marries the widowed queen Jocasta—his natural mother.   When Oedipus eventually learns that the truth of his situation, he puts out his eyes.

Freud, however, infers that Oedipus unconsciously desired to take his father's place and to have intercourse with his mother, as evidenced by his dramatic exit from Corinth. Freud extended this inference even further in assuming that Oedipus' "wish" was experienced by all males during early psychological development. According to Freud's (1905b/1953, 1925/1953) theory of psychosexual development, a boy develops an erotic bond with his mother, assuming that she too has a penis. Father is viewed as a rival for mother's affection, and the boy fears that in a jealous rage father may castrate him. At the same time, the boy recognizes that mother does not have a penis, and she becomes a reminder or symbol of this horrible penisless state of being. As the boy realizes his separateness from mother and similarity to father, he avoids castration anxiety by identifying with father and severing erotic ties with mother. Father's masculinity is incorporated as the boy's superego. After puberty, the boy re-creates his original erotic relationship with mother by choosing female love-objects that resemble her. However, according to Freud, adult male heterosexuality is never quite free of psychic conflict and rides atop a roiling river of neurotic castration anxiety. Unconscious erotic conflicts erupt from time to time for men as sexual dysfunction such as premature ejaculation or impotence.

Freud's theory of female (hetero)sexual development was also derived from Greek myth and, again, he believed that the story represented an intrapsychic conflict experienced by all girls (Freud, 1905a/1953, 1925/1953). In the Greek tragedy, Electra is deeply devoted to her father Agamemnon, the great king of Mycenae, who fights in the long war against Troy. During his protracted absence, Clytemnestra, Electra's mother and the queen, rules the kingdom and enjoys a passionate affair with Aegisthus. Electra, aware of the affair and of her mother's betrayal of her father, grows to hate her mother. When the war in Troy ends, King Agamemnon returns triumphantly but is soon murdered in his bath by his wife, Clytemnestra. The queen, fearing and despising Electra, gives her in marriage to a poor farmer who lives far away from the city. Eventually, Electra's brother Orestes avenges his father's murder by killing both Aegisthus and his mother Clytemnestra.

In adapting the Electra story to female psychosexual development, Freud took an even more circuitous and peripheral route than he did with the Oedipus myth; the theory hardly resembles the myth at all. Freud (1905a/1953, 1925/1953) speculated that a girl's first love object is her mother, her caretaker. The girl soon recognizes that she differs physically from father and that he has a penis, while mother does not. The girl desires to have a penis too and blames mother for her penisless condition. The girl shifts her erotic attachment to father who has the envied phallus. At the same time, she identifies with mother who shares her penisless state. Freud believe that this tension between identification with mother and love of father lasts much longer and is less likely to be resolved for girls than is the Oedipal complex for boys. The girl manages this conflict by *sublimating* her erotic attachment to father. After puberty, the girl unconsciously seeks male love objects that resemble father in order to re-experience this early erotic relationship. Because these unconscious conflicts are never completely resolved, they erupt from time to time for women as sexual dysfunction, including lack of sexual desire.

Despite his complex and comprehensive theory of "normal" psychosexual development, Freud, unlike his followers, never presented other-sex erotic attraction as inborn (Lewes, 1988). For Freud, other-sex erotic attraction is not inherent but is a consequence of resolving particular intrapsychic conflicts. Thus, heterosexuality, while *normal*, is not *natural* in Freud's mind because it does not occur automatically (Katz, 1995).

Unfortunately, Freud's many ideas about the development of sexual attraction are more philosophical than scientific and as such are largely untestable by empirical methods (Fisher & Greenberg, 1977; Lewes, 1988). His theories are often circular and based on single-case clinical studies. Where his theories are testable, in the case of same-sex erotic development, they are not substantiated much of the time. Moreover, Freud's erotic theories are predicated on the veracity of his stages of psychosexual development. Yet Goldman and Goldman (1982) in their seminal study of children's sexual thoughts found little evidence that Freud's psychosexual stages exist. Rather, children's sexual thinking tends to conform to Piaget's stages of cognitive development. In brief, Freud's theories on the development of sexual attraction are not supported by empirical data.

In spite of the lack of empirical support for Freud's theories on sexual attraction, psychoanalytically trained cultural anthropologists up through the late 20th century claimed to have found symbolic evidence of Oedipal conflicts and sexual neuroses among a variety of non-Western cultures, although much of this evidence is ambiguous or casts doubt on psychosexual theory [on the Sambia, see Herdt (1982) and Stoller and Herdt (1985); on the Gebusi, see Knauft (1987); on the Orokaiva-Binandere, see Schwimmer (1984/1993)]. It is doubtful that the Western middle-class Victorian values and psychic conflicts over sexual desires inherent within Freudian theory can tell us much of importance about non-industrial, non-Western cultures. Applying Freudian psychoanalytic theory to pre-industrial, non-Western cultures without examining the theory's assumptions is both ahistorical and culturally patronizing. That is not to say, of course, that cultural symbols are unrelated to social anxiety or sexual beliefs. But, attempts to decipher a culture's symbols using Western psychoanalytic theory as the gold standard leads in the wrong direction—away from native meaning and symbols from the culture itself.

2.1.2 Post-Freudian Psychoanalytic Theories

After Freud's death, his followers sharply diverged from his core concepts. Post-Freudian psychoanalysts, for instance, produced several theories of sexual attraction that decisively viewed male same-sex eroticism as pathological (Lewes, 1988). For 40 years after Freud's death, psychoanalytic theories of male same-sex erotic attraction became increasingly moralistic in tone and reflective of conventional social prejudices. Contrary to Freud, his followers located the "cause" of male-male eroticism in the earlier, oral stage of psychosexual development or thought it resulted from dysfunctional family dynamics. Post-Freudian psychoanalysts were so sure of their assumptions that they made little pretense at being scientific. As a result, numerous psychoanalysts drew their samples of male "homosexuals" from the criminal population and the most severely disturbed psychiatric patients. Post-Freudians also summarily rejected Freud's central theorem of universal bisexuality and instead "naturalized" other-sex erotic attraction (Bergler, 1947; Rado, 1940). Psychoanalysts then claimed that heterosexuality was inborn and "normal," while (male) homosexuality was neither; being *natural* and *normal*, other-sex attraction needed no further explanation. Although Freud himself used the term *normal sexuality* to refer to male-female erotic attraction, he did not believe that other-sex attraction was inherent or that it developed inevitably. In contrast, post-Freudian psychoanalysts, however, used the term *normal* to indicate that other-sex erotic attraction was inherent and standard; all other forms of sexual attraction were perversions. Consequently, psychoanalysts viewed male same-sex eroticism as problematic, requiring an explanation and cure. Theories of male-male eroticism flourished during this period of time, as did therapies designed to cure male homosexuality.

One of the more widespread psychoanalytic theories of same-sex male erotic attraction came from a 1962 study by psychiatrist Irving Bieber and several associates. Bieber and colleagues surveyed the analysts of 106 severely disturbed "homosexual" patients and the analysts of an unmatched sample of 100 "heterosexual" patients. The patients themselves were *not* surveyed. Essentially, Bieber collected psychoanalytic interpretations about disturbed "homosexual" patients who were in clinical treatment. He found that not one "homosexual" patient had a warm and caring relationship with his father, and many "homosexual" patients had "close-binding-intimate mothers" who were "seductive" with their sons. Consequently, Bieber concluded that same-sex erotic attraction in men resulted from dysfunctional family dynamics—more precisely, same-sex male attraction was the product of a dominant and overprotective mother and a passive and rejecting father (Bieber, 1976). Nearly 10 years later, British psychiatrist Charlotte Wolff (1971) drew similar conclusions about the development of erotic attraction among "lesbians." Wolff examined more than 100 nonpatient "lesbians" and compared them with a control group of "heterosexual" women matched for social class, occupation, and family background—a vast improvement over Bieber's methodology. Wolff found that "lesbians" were more likely than "heterosexual" women to have a rejecting or indifferent mother and a detached or absent father. She speculated that because of their dysfunctional home life "lesbians" failed to learn how to relate to men and attempted to gain their mother's love from other women. Both Bieber and Wolff concluded that the obvious treatment for same-sex eroticism was to correct these flawed family dynamics. Although Bieber simply ignored the inherent methodological biases of his sample, neither Bieber nor Wolff considered alternative explanations for their findings. Assuming for the moment that Bieber and Wolff have valid data, both studies found only correlations, which are not to be mistaken for causal relationships. Correlational relationships can operate in the reverse direction. That is, parents may have reacted to uncomfortable emotional or behavioral tendencies in their "gay" children by pulling away from them. Rather than a cause, the poor family dynamics of "homosexual" patients reported in these studies may have been an effect of parenting children with same-sex erotic affections. More systematic and reliable research has not substantiated the claim that parents parent "gay" children any differently than they parent their "heterosexual" children (Isay, 1989; Ross & Arrindell, 1988).

Other psychoanalysts, based on clinical case studies, attributed same-sex male eroticism to oral fixation and a failed separation from mother

(Socarides, 1968; Socarides & Volkan, 1990). These theories enjoyed wide popularity for a time. Edmund Bergler (1947), while accepting that male-male sexual attraction resulted from intrapsychic conflicts, also purported that same-sex male eroticism could be learned accidentally. In other words, anyone could become "homosexual." The idea that homosexuality could occur accidentally to anyone evoked a great deal of fear. Bergler, as well, perpetuated the myth of "homosexual" seduction and warned parents and the public to beware of "homosexual predators" because "your teen-age children may be the victims" (Lewes, 1988, p. 153). The fear-mongering Bergler built his career on the "study" of same-sex male eroticism. He was so convinced of the severe psychopathology of homosexuality that he stridently declared, "There are no happy homosexuals," despite his own patients' claims to the contrary (Lewes, 1988, p. 131).

The constant theme of "homosexual" dysfunction by post-Freudian psychoanalysts flew in the face of disconfirming evidence from patients, ordinary citizens, and empirical studies. Obviously, people with same-sex erotic sensibilities were not always mentally ill or unhealthy. At the same time, other-sex erotic attraction was no guarantee of mental health or happiness. In the 40 years after Freud's death, psychoanalysts attributed all sorts of psychopathology to the social stigma associated with homosexuality that had nothing to do with science. In a clever study that drove home this point, psychiatrist Evelyn Hooker (1957) asked skilled clinicians to identify which responses on a Rorschach ink blot test came from male "homosexuals" and which ones came from "heterosexuals." At the time, the psychiatric community viewed the Rorschach as a reliable instrument for diagnosing male homosexuality. When the results were tallied—much to the dismay of leading psychoanalysts—study clinicians performed no better than chance at identifying men with same-sex desires. Hooker concluded that there was no valid homosexual profile in Rorschach responses and that "gay" men were no more pathological than were "heterosexual" men. Despite data to the contrary, prominent psychoanalysts Edmund Bergler and Richard Socarides were unmoved in their conviction that homosexuality is inherently pathological. They soundly dismissed empirical studies like Hooker's as unrepresentative of the "homosexual" population and inconsequential to their more refined and experienced clinical "wisdom" (Bayer, 1987; Lewes, 1988).

Apart from ignoring contradictory data, perhaps the most damning evidence against post-Freudian psychoanalytic theories of same-sex male eroticism was their failure to find a cure. Although a few psychoanalysts such as Bergler (1947) and Socarides (1969) boasted of having impressive cure rates, they presented no systematic verifiable body of evidence to support their claims. Other studies noted that nearly all patients who underwent psychoanalysis to cure their same-sex desires were disappointed with the results (Lewes, 1988). In looking back at the four decades following Freud's death, the ease with which negative cultural beliefs about male-male eroticism were appropriated by psychoanalysts (and other mental health and health-care professionals) and repackaged as scientific fact is astounding. Unequivocal "evidence" of psychopathology began with the unfounded assumption that same-sex male eroticism was an intrapsychic gender disturbance (Lewes, 1988). If psychoanalysts had assumed that other-sex attraction was pathological, they no doubt would have discovered considerable evidence of mental illness among "heterosexuals." It is worth pointing out that the pious insistence by post-Freudian psychoanalysts on their unquestionable clinical wisdom and rejection of empirical methodology are quite similar in form and practice to the approach taken by the Christian Right on the issue of same-sex eroticism. It is also tempting to view early male psychoanalysts' vehement pathologicalization (and fear) of male same-sex eroticism—particularly statements by Bergler and Socarides—within the context of their theories on latent or sublimated homosexuality. However, without evidence of their private desires, it is more parsimonious simply to view Bergler and Socarides as both products of a heterosexist/homonegative culture and supporters of a like-minded theory.

Overreliance on clinical case histories of severely disturbed patients, rejection of empirical methodology, and intolerance of alternative explanations or criticism of their methods contributed to the discrediting of psychoanalytic theory in the late 1960s and 1970s. In the early part of the 1970s, psychoanalysts' unswerving belief in their judgment and in the pathology of same-sex eroticism suffered a direct assault from data-wielding, angry "gay" activists who insisted on speaking on their own behalf. In 1973, after many impassioned arguments from both sides, the American Psychiatric Association removed "homosexuality" as a diagnosis from the official compendium of diagnoses: *The Diagnostic and Statistical Manual of Mental Disorders*. Ronald Bayer, in his 1987 book *Homosexuality and American Psychiatry*, presents an entertaining behind-the-scenes narrative of the process leading to the deletion of "homosexuality" as a mental disorder.[2] Several psychoanalysts fought hard to retain "homosexuality" as a mental illness and feared that the removal of "homosexuality" as a diagnosis would contribute to the loss of their authority and, thereby, weakened the profession.

Soon after the American Psychiatric Association dropped "homosexuality" as a pathology, two other mental health bodies—the American Psychological Association and the National Association of Social Workers—followed suit (Bayer, 1987). Officially, homosexuality was no longer a mental illness. However, for many people, mental health professionals and the general public alike, strong negative attitudes about same-sex eroticism remained.

The subsequent emergence of feminist psychoanalytic theory in the mid-1970s and the more recent predominance of biologically based sexuality research have contributed to a reformulation of psychoanalytic theories of sexual attraction. Reformed psychoanalytic theories are better grounded in extensive clinical observations and empirical data, and they do not view same-sex attraction as a problem demanding a cure (Friedman, 1988; Isay, 1989). Psychiatrist Richard Isay, for example, suggests that boys who grow up to be "gay" assume cross-gender characteristics such as gentleness and sensitivity as a means to obtain attention and affection from their father. Boys who grow up to be "heterosexual" develop gender-typical characteristics for similar reasons—to attract mother's attention and, later, to attract someone like mother. Isay hypothesizes that "gay" men distance themselves from their father in their memory to avoid recognition of their erotic attachment to him. On the other hand, father may also withdraw from a "gay" son when he becomes aware that his son does not act like other boys his age or in reaction to his anxiety about his son's attachment to him. The objective of reformed psychoanalysts in providing psychotherapy to "gay" men and "lesbians" is to help the patient recognize these erotic undercurrents and find adaptive, healthy ways to live—not to change their sexual orientation.

2.1.3 Post-Freudian Theories and Non-Western Cultures

From an empirical point of view, post-Freudian psychoanalytic theories of same-sex eroticism have not found support. Post-Freudian psychoanalysts have no general theory of sexual attraction; they have only loosely related theories of same-sex male erotic attraction. Kenneth Lewes (1988), in his thorough review of psychoanalytic theories of same-sex male eroticism, described post-Freudian theories of sexuality as circular, antiempirical, and *gyneco*phobic. This last charge may help explain psychoanalysts' obsession with cross-gender characteristics of men with same-sex desire and their relative disinterest in female sexuality. Beginning in the mid-1970s, female psychoanalytic writers inspired by the feminist movement explored female sexuality from a positive point of view.

Although contemporary psychoanalytic theorists do not view same-sex attraction as dysfunctional and employ empirical methodology in their research, applying a culturally specific theory like psychoanalysis to non-Western cultures presents several problems. First, historically and in other cultures, there is little objective evidence to support psychoanalytic conceptions of sexual attraction, although many psychoanalytically trained anthropologists have tried. Second, the arrogance of claiming to know secrets of an alien culture is obvious, and the probability of error is extremely high that one can decipher non-Western cultural symbols from an urban, middle-class, sex-negative, psychoanalytic perspective. The bottom line is that it is just not possible to know with any degree of certainty that same-sex male erotic attraction among non-Western people is related to unconscious emotional conflicts, desire for father's affection, or even dysfunctional families.

Evidence against psychoanalytic explanations, however, is easier to find. For example, psychoanalytic theory supposes that same-sex male eroticism is related to extremely close mother relationships, cross-sex identification, cross-gender characteristics, and sexual immaturity and that the theory is universal. But, there is nothing to suggest that the ancient Babylonian Gilgamesh or his companion, the barbarian Enkidu, were enmeshed with their mothers or effeminate men. They referred to each other as husband and wife, and Gilgamesh wept "like a wailing woman" at Enkidu's death (Halperin, 1990, p. 81). This male couple simply reflected the social expectations of their culture and class (Greenberg, 1988). We know that Gilgamesh did not confine his sexual activity to one sex. In all, the story of Gilgamesh and Enkidu contradicts what psychoanalytic theory would predict about each man.

Could we say that David, the future king of Israel, and Jonathan, Saul's son, were the products of dysfunctional families? Certainly, Saul's dysfunction as an individual and as a parent is easy to observe. But, did Jonathan have an unusually close relationship with his mother? Was Jonathan effeminate because he "made a covenant [with David], because he loved him [David] as his own soul" (1 Sam. 18:3)? Was David effeminate because Jonathan loved him? Both men were brave soldiers and close in age. Without a doubt, Jonathan and David cared deeply for each other, and when Jonathan died, David declared that Jonathan's "love to me was wonderful, passing the love of women" (2 Sam. 1:26). Again, the story of David and Jonathan fail to fit the psychoanalytic model.

The legendary Greek male warrior couple, Achilles and Patroklos, were hardly effeminate men, although there was some confusion at the time over which social role each played—whether Achilles was *erastēs* (lover) or *erōmenos* (beloved) of Patroklos (Cantarella, 1992; Dover, 1978/1989; Halperin, 1990). Love between males in ancient Greece, like that between males and females, was expected to be hierarchical, usually based on age, with the older man loving a male youth or girl. In reality, however, the ideal of generational difference may have been frequently violated (Greenberg, 1988; Halperin, 1990). So, perhaps, Achilles and Patroklos did not strictly follow social custom in their relationship. Were they sexually immature? Is that why they chose a male partner? The Achaeans believed that male-male love was superior to loving women. Achilles' relationship with Patroklos, however, did not prevent him from having sex with female slaves. Still, every male was expected to marry. It is somewhat surprising then to hear Achilles' mother, the goddess Thetis, counsel him to take a wife "as is proper" after mourning the death of Patroklos (Hom. *Iliad* XXIV). Apparently, Achilles and Patroklos too, both grown men, had postponed marriage for some time. Although this may have been unusual, there is no evidence from Homer's story that the culture viewed Achilles and Patroklos as pathological.

In most ancient societies, sexual relationships were structured by social status such as *age* or *sexual role*—active/insertive versus passive/receptive—and not by sexual identity. In general, most ancient societies did not view same-sex male attraction itself as a problem or pathology, contrary to psychoanalytic theory. However, by the Greek Hellenistic period, male effeminacy was associated with sexual receptivity, which was increasingly stigmatized among Greeks and Romans and by Western culture up to the present (Greenberg, 1988). Yet playing the active sexual role in a male-male relationship bore no stigma.

For the sake of space, I will give only one more example of the poor fit of psychoanalytic theory with pre-industrial cultures. Among some Native American Indian societies, men and women who experience same-sex erotic attraction represented a *third gender*, which represents an intermediate position between men and women. Such individuals possessed both male and female characteristics and were called *berdache* (those with male physiology) or *amazon* (those with female physiology) (W. Williams, 1986). The term *berdache*, which loosely translates as sodomite, is stigmatizing and is not descriptive (D'Emilio & Freedman, 1988/1997; Greenberg, 1988). The more recent and neutral term *two-spirited people* is preferred. Several Native American Indian nations had three genders and *two-spirited people*, including the Iroquois, California, Eskimo, Comanche, Cherokee, Illinois, Nadowessi, Chippewas, Koniag, Oglala, Quinault, Crow, Cheyenne, Creek, Yokot, Sioux, Fox, Sack, Zuni, Pima, Mohave, Navaho, Cree, Dakota, Siksika, Arikara, Mandan, Florida, and Yucatan (Bleys, 1995; Greenberg, 1988; W. Williams, 1986). A pre-adolescent boy or girl made their choice of gender known to the community through public interpretation of dreams, choice of dress, or choice of sex-typed objects during a ritual. Male *two-spirited people* wore at least some articles of female clothing, used female mannerisms and speech patterns, and engaged in female-typical occupations. Most male *two-spirited people* had sexual relations exclusively with married men in the community and, sometimes, they married men themselves. In other Indian nations, *two-*

spirited people were celibate or had sexual relations with both men and women but could not marry each other because to do so would be a selfish misuse of their spiritual gifts. In many North American Indian nations, male and female *two-spirited people* were respected as shaman or lucky individuals who acted as spiritual intercessors for members of the community and counselor between men and women.

Although female *two-spirited people* were less common (or just less often the subject of inquiry), they too generally held positions of authority within the community. Female *two-spirited people* could be found among the Kutenai, Snake, Cocopa, Klamath, Crow, Mohave, Maricopa, and Kaska of American Indian nations (Blackwood, 1993; Greenberg, 1988). However, from the moment that European explorers and missionaries came in contact with American Indians, the institution of *two-spirited people* began to erode. Today the concept of *two-spirited people* has almost disappeared among American Indian communities, and Indians who have same-sex erotic feelings are more likely to identify as "gay" or "lesbian" than as *two-spirited people* (W. Williams, 1986).

Back to the question of whether psychoanalytic theory applies to the development of sexual attraction among American Indians. Nowhere in the descriptions of *berdache, amazon,* or *two-spirited people* is there verification of oedipal conflicts, sexual anxiety, or overidentification with mother, although *two-spirited people* were certainly androgynous or cross-sexed in dress and mannerism. Yet, it is unfair to make such a comparison. The three-gender system practiced by many American Indian nations is not the same as the two-gender system practiced by white European culture. What is more, among Indian cultures that recognized *two-spirited people*, the individual chose their gender during a community ceremony rather than have a gender assigned to them by authority figures, as in European culture. And although some American Indian cultures did not hold *two-spirited people* in high regard, same-sex erotic attraction and behavior was not stigmatized, and married men still enjoyed sexual encounters with male *two-spirited people* (Greenberg, 1988; W. Williams, 1986). In short, psychoanalytic theory does not provide a good explanation for the development of sexual attraction in American Indian cultures, or in non-Western, pre-industrial cultures, for that matter.

2.2 LEARNING AND CONDITIONING

2.2.1 Classical Conditioning

A second major psychosocial explanation for the acquisition of sexual attraction comes from behavioral *learning theory*, specifically, from the principles of *classical* and *operant conditioning.* Joseph Wolpe (1958) was one of the first behaviorists to apply conditioning principles to sexual attraction—or, more precisely, to sexual *paraphilias.* Wolpe suggested that sexual disorders might result from *faulty* learning. The implication was that typical sexual attraction was a product of *appropriate* learning. During the 1960s and early 1970s, behavioral theories of sexual attraction were quite popular, although, like most other sexuality theories, they focused mainly on male sexuality and on same-sex male eroticism in particular. Behavioral theories of male same-sex eroticism were widely accepted for a time in part because of the lack of scientific precision of psychoanalytic theory but perhaps also because parents of "gay" children were eager to divest themselves of the blame and responsibility dumped on them by psychoanalysts. Behavioral theories, by contrast, claimed that individuals were "gay" because they *learned* to be that way. While strict behavioral learning theories of same-sex erotic attraction have little support today, variants of these theories flourish in the form of *sexual orientation conversion therapies* or so-called *reparative psychotherapy* among "ex-gay" organizations. *Ex-gay* refers to someone who was once "gay" but now identifies as "heterosexual" as a result of religious conviction or *reparative* psychotherapy. The "ex-gay" ministries use this term, although there is no reliable evidence that any recipient of *reparative* treatment changed their sexual orientation. Reported "cures" usually mean not acting on same-sex desires, marrying someone of the other-sex, and living an outwardly "heterosexual" lifestyle. Some "ex-gays" and "ex-ex-gays" acknowledge that same-sex erotic desire did not disappear as a result of treatment; "ex-gays" prefer to emphasize their desire for a family and for a socially accepted lifestyle (Paulk, 1998; Pennington, 1989). "Ex-ex-gays" are people who at one time claimed to have converted to heterosexuality but now, again, identify as "gay."

Learning theories utilize some form of either classical or operant conditioning, which are quite distinct. In brief, *classical conditioning* holds that an *unconditioned stimulus (UCS)*—the presentation of food—evokes a reflexive behavior—an *unconditioned response (UCR),* such as salivation in dogs, to borrow Pavlov's famous example. When a *neutral stimulus (NS)*—a bell—is paired with the UCS, after a time, the NS takes on the stimulus properties of the UCS and becomes a *conditioned stimulus (CS)*. Now, when the CS—the bell—is presented alone, a *conditioned response (CR)*—salivation—occurs. The CR is not identical to the UCR, but varies in quality and strength from the UCR. Classical learning theorists presume that most behaviors are conditioned or learned in this way.

To apply this principle to the development of same-sex male erotic attraction, when a UCS—genital stimulation—is paired frequently with a NS—presence of another male—the NS becomes a CS and alone evokes a CR—sexual arousal or erection. Thereafter, males evoke sexual arousal for this man. While classical theorists usual portray the learning of same-sex male eroticism as accidental, this must be a frequent or potent accident in order for the association to carry. To *extinguish* a classically conditioned response, the CS is presented alone, until it no longer evokes the CR; thus, breaking the learned association. In our example of same-sex erotic conditioning, males—the CS—would be presented until no sexual arousal occurs for the subject in the presence of male stimuli.

Classical conditioning theorists never describe the learning sequence for other-sex erotic attraction. By this significant omission, conditioning theorists imply that other-sex attraction is a reflexive behavior that somehow develops *naturally*. On the other hand, classical conditioning theorists present same-sex male eroticism as a problem and an *error* of learning as a result of the mistaken pairing of male stimuli with an UCS for sexual arousal.

In one of the first studies of classically conditioned sexual arousal, Rachman (1966; Rachman & Hodgson, 1968) demonstrated that non-sexual stimuli could elicit sexual arousal and perhaps explain the development of sexual fetishes. Rachman paired pictures of knee-high black women's boots (NS) with photographs of nude women (presented as an UCS in the study but actually a CS)[3] until the eight ("heterosexual") men in his study got an erection, as measured by a penile plethysmograph, when shown the boots alone. Although his results were mixed and not definitive, Rachman concluded nonetheless that sexual fetishes could be explained by and treated with conditioning principles.

Rachman's conclusions were overly optimistic. First, a photograph of nude women, as a representation of live women, is actually a CS, not an UCS. The investigators also assumed that sexual arousal conditioning had already occurred to females for the men in these studies. In other words, Rachman attempted to condition arousal using a CS, a photograph of nude women, which, being somewhat removed from the original UCS, would have fewer stimulus properties. If he had paired nude women with black boots, his results might have been much stronger. Second, a closer reading of Rachman's studies reveals that subjects exhibited a CR to the CS—erection to the boots alone—*only after 24 to 65 pairings!* Learning was highly variable in these studies, and at 65 pairings, boot fetishism would not seem to be an easy thing for men to learn. In real life, how likely is it that non-sexual objects are paired with sexual arousal as many as 65 times? More damning, however, to Rachman's argument, he dropped two subjects from his study when they failed to experience an erection after *more than 60 presentations* of the UCS-NS pair, and he eliminated one subject because he showed *almost no sexual arousal to any stimuli* that were presented! Obviously, some men are not susceptible to sexual conditioning of a boot fetish, at least in the manner devised by Rachman.   Previous reviewers have cited additional methodological problems with Rachman's findings that cast a long shadow of doubt on the study's validity and generalizability. For example, Rachman did not define the minimum penile response necessary to meet criteria for a CR, nor did he control for pseudo-conditioning by also presenting novel stimuli (Alford, Plaud, & McNair, 1995).   Both of Rachman's studies also had extremely small samples of three and five subjects each. In all, Rachman's work, while initially heralded as an explanation for the development of sexual learning, fails to provide convincing data that sexual attraction is learned in this way.

In another famous study, Nathaniel McConaghy (1967) paired red and green circles and triangles (NS) with "homosexual" and "heterosexual" film erotica as the UCS. As predicted, "gay" men developed an erection when presented with geometric shapes (CS) that had been paired with male erotica, and "heterosexual" men got an erection when presented with geometric shapes that had been paired with female erotica. Men in both groups lost their erections when the shapes were paired with their nonpreferred erotic images. In McCongahy's view, "gay" men showed a negative erotic response to images of nude women, and "heterosexual" men had a similar reaction when presented with images of nude men. He concluded that objects do acquire sexual association but that conditioning is specific to sexual orientation—which itself showed no evidence of conditioning.

However, two other behaviorists, Langevin and Martin (1975), criticized McConaghy's study for confounding natural detumescence following arousal with a negative CR to nonpreferred erotic stimuli; McConaghy did not allow sexual arousal in his subjects to return to baseline between image presentations. Langevin and Martin also faulted McConaghy for not using a random control procedure and for not using novel stimuli to test the unconditioned effects of the CS.   Still, Langevin and Martin's explanation that an expected period of detumescence interfered with or masked sexual arousal to alternative stimuli seems weak to me. If sexual attraction is pliable, as Langevin and Martin assume, it is reasonable to believe that arousal could be conditioned to most stimuli that is paired with erotica. Certainly, a period of detumescence is important methodologically to demonstrate a conditioned response to a particular stimulus and no other. However, if conditioned had occurred, both "gay" and "heterosexual" men should have maintained their erections across stimuli, although the strength of erection might vary. Given Langevin and Martin's assumption that sexual attraction can be conditioned, learning should continue beyond one conditioned response. Nonetheless, "gay" and "heterosexual" men evidence a clear preference in erotic stimuli and were not aroused by images of their nonpreferred erotic images. An explanation that was not considered for McConaghy's results is that adult sexual arousal is sex-specific, well learned, and subject to social desirability and congruence with sexual identity. Said another way, adult males respond easily to socially acceptable erotic stimuli, but arousal to stigmatized stimuli that are incongruent with sexual identity may be suppressed.

Langevin and Martin (1975) conducted their own classical conditioning experiments of sexual arousal and attempted to correct Rachman's earlier methodological problems. The authors used erotic pictures from men's magazines as their UCS and included non-sexual stimuli as the NS in paired presentations. Eventually the non-sexual pictures elicited an erectile response in (heterosexual) male subjects. When Langevin and Martin attempted to reverse the learned association, extinction occurred quickly— after only 10 trials of the CS alone. More critically, however, the smallest amplitude of penile arousal was statistically significant, rather than the frequency of response. This finding suggests that sexual arousal to the conditioned non-sexual stimuli was slight and did not increase in strength over time (Alford, Plaud, & McNair, 1995). In other words, conditioning was poor, or the response-eliciting power of the UCS (photographs of nude women) was weak. In treating photographs of nude women as comparable stimuli to nude women, Langevin and Martin repeat Rachman's error and weaken potential conditioning. Although it seems unlikely that photographs of naked women would not be sexually evocative enough for "heterosexual" men to condition to non-sexual stimuli, alternative explanations are that stimuli are not equivalent and conditioning is selective or that sexual arousal is not conditioned in this way. For other critical reviews of sexual conditioning studies, see Alford, Plaud, and McNair (1995) and Diamant and McAnulty (1995).

Some behaviorists have argued that *natural selection* favors sexual associations to particular stimuli (Garcia & Koelling, 1966; Seligman, 1970). In other words, men and women may be biologically prepared to make sexual attributions to select stimuli but not to all. Sex might be one of those characteristics that we are predisposed to favor. Thus, the sex to which we are attracted would be largely consistent over time and resistant to change.

One sex might also be biologically prepared to make sexual associations more often than the other sex. This would explain why men more than women sexualize objects. In addition, we may be biologically prepared to sexualize some stimuli more than others; brassieres, panties, lingerie, body hair, body scent, and so forth are more often sexualized than common objects such as pillows, bed sheets, beds, or ceilings which are found in environments where sex often occurs.

Although we know little about sexual conditioning of men, even less is known about sexual conditioning of women. For the most part, studies of sexual conditioning have ignored women and have focused on male sexual paraphilias and re-conditioning "homosexual" men. These emphases point out which aspects of sexuality were of particular interest to behaviorists and to the larger culture.

2.2.2 Operant Conditioning

Unlike classical conditioning, *operant conditioning* holds that a consequence that follows a behavior influences the likelihood that the behavior will occur again. In the case of sexuality, if the consequence leads to the increased frequency of the sexual behavior of interest, then the behavior is *positively reinforced*, and the consequence is a *positive reinforcer*. Orgasm, for example, may be a positive reinforcer for sexual intercourse if intercourse occurs more frequently as a result of experiencing an orgasm. If, however, avoidance or escape from a consequence results in the increased frequency of the sexual behavior of interest, the behavior is *negatively reinforced*, and the threat is a *negative reinforcer*. By avoiding the threat of disease or pregnancy, condom use, for instance, may be negatively reinforced. Other behaviors are both positively *and* negatively reinforced, which make them especially resistant to change. Masturbation, for example, may be positively and negatively reinforced, if masturbation is more likely to occur as a result of experiencing tension release and pleasurable feelings following orgasm. Furthermore, a *discriminative stimulus* (S^D)—soft lights, romantic music, a smiling companion—often signals the presence of positive or negative reinforcers in the environment. For example, coming home to a candlelight dinner, soft music, and a smiling spouse may be strong S^Ds for positive reinforcement of sexual behavior.

On the other hand, consequences that lead to the decreased likelihood of the behavior of interest are *punishers*. Receiving an electric shock after or at the point of orgasm during masturbation may significantly reduce the likelihood of the sexual behavior in the future. If the individual masturbates less often as a result of the shock, then punishment occurred. If, however, masturbation continues at about the same frequency, then the shock was not a punisher.

Like their classical counterparts, operant conditioning theorists have generally assumed that other-sex erotic attraction is automatic or *naturally* learned. At the same time, operant theorists seem to view same-sex eroticism as a product of inadvertent or faulty learning, a process that demands explanation. Operant conditioning theorists have speculated that same-sex erotic attraction results from seemingly innocuous experiences such as accidental stimulation of an infant's genitals by the same-sex parent, punishment by the other-sex parent following genital stimulation, negative social messages about other-sex erotic relations, positive attention from an important same-sex individual, arousal during same-sex peer play, lack of an other-sex partner when sexually aroused, or from poor heterosocial skills that interfere was "appropriate" reinforcement (Barlow & Agras, 1973; Green, 1985, 1987; Greenspoon & Lamal, 1987). No single experience results in same-sex erotic attraction but, rather, a pattern of experiences channel sexual attraction, according to operant theorists.

Part of the problem with this theory is that most people experience some or all of these early psychosexual experiences and do not grow up to be "gay." Operant conditioning theorists—Barlow and Agras (1973), Green (1985, 1987), and Greenspoon and Lamal (1987), for example—do not explain why certain common experiences are so reinforcing for a minority of individuals. On the other hand, a few researchers have noted that both "gay" and "heterosexual" adults report having had similar parenting experiences (Isay, 1989; Ross & Arrindell, 1988). Other researchers have found that early sexual contact with same-sex or other-sex persons is not predictive of adult sexual orientation (Bell, Weinberg, & Hammersmith, 1981). Furthermore, researchers have reported that many "gay" people have had a great deal of other-sex erotic experience, contrary to the supposition that they lack heterosocial skills and sexual experience (Kinsey *et al.*, 1948; Kinsey *et al.*, 1953; Bell & Weinberg, 1978). Operant theorists have difficulty explaining why "gay" people prefer same-sex erotic stimulation despite frequent other-sex experiences and why "heterosexuals" prefer other-sex stimulation despite having some same-sex erotic experiences.

A couple of learning studies allegedly provide support for the idea that same-sex eroticism is operantly conditioned. One study found that individuals who learn to masturbate by being manually stimulated by someone of the same sex and who experience their first orgasm during a same-sex contact are more likely to report having same-sex erotic feelings and an adult "gay" identity (Van Wyck & Geist, 1984). However, again, this is a correlation. It is entirely possible that same-sex erotic feelings precede sexual contact and lead to its initiation, instead of the other way around.

A second operant learning study hypothesized that boys learned to be "gay" in the absence of strong male models of behavior (Reiss, 1986). In fact, same-sex erotic behavior is more often found in male-dominant, gender role-rigid societies in which father contact with infants is low. According to this theory, infrequent contact with father does not allow boys sufficient exposure to "masculine" behavior, and so boys subsequently socialize with women and romanticize men. However, in patriarchal gender role-rigid cultures, it is inconceivable that male models of behavior are ever truly absent. Rather, I hypothesize that the high social status given to men in such societies makes "masculine" behavior more powerful and salient; it is the status of male behavior in those cultures that makes the model powerful, not direct contact hours with men. What is more, females in patriarchal gender role-rigid cultures participate in socializing children to fit the behavioral norms appropriate for their sex. Fathers are not the only source of information for boys about "masculine" behavior; mothers and other females collaborate with men in teaching gender role behavior to children. Socialization of gender roles is much more communal and pervasive than many behaviorists describe (Barlow & Agras, 1973; Green, 1985, 1987; Greenspoon & Lamal, 1987).

In a third study, sociologists John and Janice Baldwin (1989) gave an operant learning spin for the supposed switching from "homosexual" to "heterosexual" sexual orientations by Sambian males in New Guinea. To grasp the complexities of this culture and of Baldwins' hypothesis, a brief overview of the Sambia and related cultures in New Guinea is necessary.

Sambia is not the actual name of this Melanesian community. Anthropologist Gilbert Herdt (1981) penned this pseudonym in order to protect their identity from the Western world; most of what we know about the Sambia comes from Herdt's direct observations and informant interviews over many years. The Sambia are one of several cultures, collectively known as Melanesia, which inhabit the islands of the South Pacific. Between 700 and 1,000 different cultures, sharing over 2,000 distinct languages, share these islands. The Sambia and, perhaps, 10 to 20% of Melanesian cultures engage in so-called *male insemination rituals* (Herdt, 1984/1993a). These rituals include an exchange of semen from adult males to young boys or youths. (The Sambian form of male insemination rituals is described in more detail below). Semen transactions are designed to "grow" strong virile men from boys because males are assumed not to be inherently fertile. The method of semen exchange varies across communities, from ritualized fellatio to anal intercourse. The significance of semen exchange—growing men— overshadows the sexual method of exchange. In other words, the *sexual* component of boy-insemination may be small and difficult to separate out from the larger practice of men making men from boys. At the same time, ritualized fellatio and anal intercourse between males in Melanesia are not devoid of explicit eroticism (Herdt, 1984/1993a,b). That topic is discussed later in this section.

At one time several Melanesian cultures besides the Sambia had boy-inseminating rituals, including the Kanum, Yei-anim, Marind-anim, Jaquai, Asmat, Casuarina Coast, Humboldt Bay, and Kimam of Frederick Hendrick Island of Irian Jaya; Kiwai Island, Bugilai, Keraki, Suki, Boadzi, Bedamini, Etoro, Kaluli, Onabasalu, Gebusi, Baruya, Jeghuje, Kamula, Ai'i, and other Anga groups of Papua New Guinea; Fiji, New Caledonia, the Vao and Big Nambas of Malekula Island in Vanuatu (New Hebrides), Tolai of Gazelle Peninsula of New Britain Island, Duke-of-Yorks, and East Bay of Santa Cruz Islands (Greenberg, 1988; Herdt, 1984/1993a). Because more is known about boy-insemination rituals among the Sambia than other Melanesian cultures, they will serve as a model for this particular rite.

The Sambia are a warrior people who emphasize the importance of masculine strength, the essence of which is *kweikoonbooku* or semen (Herdt, 1981, 1987). Males are not thought to naturally possess *kweikoonbooku*, and boys must receive it from older (fertile) males in order to grow and become men. Around age seven to ten, Sambian boys move into an all-male lodge with other older boys and young men who have not reached adult status. Sambians live in a sex-segregated society, where women are viewed as inferior to men. For years, boys live in an all-male enclave and avoid contact with females, including their mothers and female kin. During the next 10 to 15 years, Sambian boys participate in a series of initiation rituals. Nearly every night, boys fellate to orgasm bachelors or newly married men and swallow their semen. The accumulation of semen in the body is thought to promote physical growth and fertility. After puberty, the now-fertile youths switch roles and are fellated by younger boys, giving the younger boys the same growth-enhancing substance. Young men continue to be fellated by boys through their first few years of marriage. In Sambian culture, full manhood is not achieved until the birth of a man's second child, after which time he stops being fellated by boys and only engages in intercourse with his wife. Sambian wives are generally younger than their husbands and, while females are believed to be naturally fertile, females are also thought to require semen—obtained during sexual intercourse—to reach physical maturity. The inferior status of women in Sambia, the role of a wife as sexual partner, and the need for children for males to achieve manhood create an uneasy tension between males and females.

Back to the Baldwins' (1989) argument. They claim that Sambian boys do not enjoy performing fellatio or being fellated. They believe that

Sambian boys are coerced or forced to engage in same-sex erotic activity. Evidence for this belief is that some boys resist sucking a penis, and at the earliest opportunity, stop engaging in male-male sexual behavior. Thus, a Sambian male's first truly pleasurable, and hence, reinforcing sexual experience is with his wife, according to the Baldwins. Unfortunately, the Baldwins have misread Herdt's observations. Herdt himself discounted a learning explanation of Sambian male eroticism (Stoller & Herdt, 1985). The Baldwins also make several questionable assumptions about what Sambian boys find pleasurable and why they participate in ritual fellatio, all of which infer cognitive mediators and Western values about sexuality. The Baldwins' assertions that Sambian boys find no pleasure in giving or receiving fellatio, choose to stop male-male erotic activity as soon as possible, and only truly experience sexual pleasure with their wife reveals their Western heterosexism and the assumption that Sambian boys already have a "heterosexual" orientation. Yet, learning theories generally deny that desire or feelings initiate behavior and accept that sexual pleasure is not exclusive to male-female erotic activities. As noted in chapter 1, Western observers must be extremely cautious in deciding what is *sexual* or *pleasurable* for people from another culture or time, as well as what constitutes *volition* for people who do not share the same ideas about sexuality, personhood, or free will. Lastly, Herdt (1981, 1984/1993a, 1987) provided a wealth of information that contradicts the Baldwins' argument—most significantly, that only a *small* number of boys evidence any resistance to performing fellatio, whereas another small group readily take to it. Herdt estimated that about 5% of Sambian boys showed either of these extreme reactions regarding fellatio. What is more, Herdt believed that given fewer social constraints Sambian men would continue to engage in male fellatio throughout adulthood.

A thorough reading of Herdt's work (1981, 1984/1993a, 1987; Herdt & Stoller, 1989; Stoller & Herdt, 1985) provides a more complete explanation of male insemination rituals in Sambia. First, Sambian relationships are always age-dissimilar, involving an older male with either a younger male insemination partner or a female wife. Second, boy-insemination rituals serve erotic and non-erotic functions. For the boys, the purpose of semen exchange is to obtain *kweikoonbooku* and grow to be men. For the men, the primary purpose of semen exchange is to help boys grow physically and to socialize them as men. However, the objective for fellated Sambian bachelors and married men is their own sexual pleasure and orgasm in a boy's mouth. The insemination ritual is not always about making boys into men. Sambian men sexualize the mouth of boys and develop sexual fantasies about the mouths of particularly handsome boys (Herdt, 1981; Herdt & Stoller, 1989). Boys who are beautiful, willing, and sexually aggressive are particularly prized as fellators. Similarly, Sambian men sexualize the mouths of females, as well as other body parts. Other-sex erotic attraction among Sambian men is fairly evident, although again male-female intimate relationships embody a tension and distrust that is not apparent in male-male relationships. For this reason, male-male erotic encounters are qualitatively different than male-female relationships.

In direct contradiction to the Baldwins' argument, many Sambian men report that boys are more sexually exciting to them than are their wives (Herdt & Stoller, 1989). Herdt and Stoller described an adult man named Galako who developed an intense romantic attachment to one particular boy. Another man named Moondi experienced vivid sexual fantasies about one boy who regularly fellated him until his betrothal. After their engagement, Moondi fantasized about his fiancé. What is more, in spite of social pressure, a few Sambian men do not stop being fellated by boys after reaching manhood (the birth of their second child).

For Sambian boys, their virility hinges on their ingestion of semen. Herdt (1981) noted that while some Sambian boys initially resist performing fellatio, other boys are enthusiastic in their duties. According to custom, boys are not supposed to be sexually excited or reach orgasm while performing fellatio (Herdt & Stoller, 1989; Herdt, 1981). However, several Sambian male informants describe experiencing markedly erotic feelings while fellating men. Some older boys report getting erections and experiencing pleasurable physical sensations while sucking a man's penis. Sambian boys also joke amongst themselves about their preference for particular men as partners and make up names for the penis of favorite bachelors (Herdt, 1981; Herdt & Stoller, 1989). Boys also debate the amount and taste of a man's semen and have sexual fantasies about particular men.

Although fellatio between near-age boys and private encounters are prohibited by custom, they seem not to be uncommon. Herdt (1981) quoted one exasperated Sambian man named Tali who lamented that "we give semen to a boy. . . . And he turns around and copulates with a younger boy. We big men copulated with him and we said: 'We completed his *moo-nungendei* (gave him milk-food from the penis), but he has gone and given his share to other boys'" (p. 235). One could argue that boys who fellate other boys simply imitated their behavior with men, and nothing more. However, Tali also worried that women in the community would discover that men "play around" with males and "turn around" (give back semen) to other men, in violation of social custom (Herdt, 1981, p. 285). It is unclear from the text whether Tali was referring to the behavior of boys with other boys or to the fact that some men fellate each other because it feels good. In either case, Tali seemed to acknowledge that semen exchange is pleasurable and not just about "growing" boys. Herdt (1984/1993a,b) also believed that boy-insemination rites in Sambia and other Melanesian cultures involved same-sex erotic attraction and arousal.

Like the Sambia, other Melanesian cultures—the Etoro, Keraki, Gebusi, Bedamini, Duke-of-York, Baruya, and Highland Anga men—also exchange semen with boys via fellatio (Herdt, 1984/1993b). Consistent with Herdt's observations, anthropologist Bruce Knauft (1985) also concluded that boy-insemination among the Gebusi may be as much about affection as duty. Other observers have noted that Keraki men prefer handsome boys for fellatio, suggesting the sexualization of some boys (Williams, 1936; Herdt, 1984/1993b). In addition, among the Gebusi and Bedamini, sexual play between near-age boys is not uncommon and considered to be the private business of those individuals (Knauft, 1985; Sørum, 1984/1993).

Boy-insemination/semen exchange practices vary among Melanesian cultures. Boys among the Kamula of the Papuan Plateau receive semen through anal intercourse (Herdt, 1984/1993b). The Marind-anim, Kimam of Frederick Hendrick Island, Jaquai, Kaluli, and East Bay Islanders also inseminate boys through anal intercourse (Boelaars, 1950; Davenport, 1977; Serpenti, 1965, 1984/1993; Baal, 1934/1966). Among the Kamula and Vanuatu (New Hebrides), man-boy pairs refer to each other as *husband* and *wife*, implying intimate affection and gender-role socialization. Among the Marind-anim, men continue to inseminate boys after marriage, and man-boy pairs sometimes result in extended couplings and jealousies, which also suggests an affectionate romantic attachment (Baal, 1934/1966). Among the Jaquai, fathers are paid for their sons' sexual services with an older man (Boelaars, 1950). If an erotic relationship develops between the man and boy, the inseminating partner gives his own daughter as bride to the boy. The inseminating Jaquai man is called *mo-e* (anus father) and the boy is known as *mo-maq* (anus son). Among the Kimam, however, male inseminators are closer in age to boy initiates (Serpenti, 1965, 1984/1993). Usually the boy's youngest maternal uncle or a cross cousin provides semen for physical growth.

There is suggestive evidence that female fertilization rituals coexist with male insemination rites among the Baruya and the Big Nambas in Vanuatu (Deacon, 1934; Lindenbaum, 1987). Lactating Baruyan mothers provide breast milk to unrelated prepubescent girls in order to promote their physical maturity (Greenberg, 1988). For the most part, however, female growth rituals have not been investigated, and the existence of parallel girl-insemination rituals is largely speculative.

To summarize this section on operant conditioning, studies have failed to demonstrate that sexual attraction is acquired through learning, although preference for certain sexual activities may be learned. Cross-cultural examples of boy-insemination rituals, however, provide a somewhat different picture. After rejecting the Baldwins' application of learning theory, there is, nonetheless, suggestive evidence that men in Melanesian cultures learn to enjoy sexual relations with males and females. Most men in Melanesian cultures participate in and enjoy fellatio with other males, and in fact, nearly all men engage in sexual activity with both sexes during their lifetime.

Contrary to conventional learning theorists, sexual attraction to males is not exclusive of attraction to females; in fact, they may represent parallel attractions. While it may be argued that Sambian males experience extensive learning trials in order to become sexually aroused by other males, there is no indication from available anthropological data that males in any Melanesian culture need to practice sexual attraction to females. Attraction to females seems to just be there. On the other hand, there are reports that married men, in making the transition from exclusively male sexual partners to occasional or exclusive female partners, sometimes have sexual performance difficulties; some men have to practice sexual activity with a female partner in order to maintain their erection and achieve orgasm. However, practicing a new sexual behavior with a new partner does not mean that sexual attraction to females was absent. In reading the observations of anthropologists in Melanesia, I am struck by how quickly married men make the transition to female sexual partners and predict that it is the mechanics of sexual activity rather than sexual attraction to females that is learned. It also seems likely that the particular social customs of these cultures allow men to experience and realize their *capacity* for enjoying sexual relations with both sexes.

Of course, enjoyment of sexual relations with males and females does not imply that Sambian men have an equal attraction to both sexes. Evidence suggests that some adult men have preferences for one sex of partner, even while enjoying sexual relations with both sexes. In addition, at an early age a few Sambian boys demonstrate a preference or nonpreference for a sex of partner that persists over time. However, nothing in the accounts of Sambian or Melanesian cultures suggests that for men sexual orientation changes from an *exclusive* sexual attraction to males to an *exclusive* attraction to females. For most Melanesian men, sexual orientation includes the capacity for attraction to both sexes.

2.2.3 Sexual Orientation Conversion Therapies

Another source of information about the veracity of learning theory when applied to the acquisition of sexual attraction comes from efforts to change it. Assuming that same-sex erotic attraction is learned through conditioning, a number of American behavioral therapists in the 1960s and 1970s attempted to cure male homosexuality with *aversion* and *conversion* psychotherapies. Typical sexual re-orientation therapies involved punishing same-sex erotic responses (erection) and rewarding (praise or orgasm) other-sex responses (Barlow & Agras, 1973; Feldman & MacCulloch, 1971; Marks & Gelder, 1967). Classical conditioning conversion approaches paired aversive stimuli—an emetic, electric shock, or ammonia—with the undesired erotic response. Although the strength of erections to same-sex erotic stimuli diminished in the laboratory, in most cases, aversive responses to same-sex erotic encounters were rare (McConaghy, 1987). This finding may actually illustrate *state-dependent* learning. In operant conditioning language, the laboratory functions as a discriminative stimulus (S^D) for punishment. Outside of the laboratory, the S^D is absent and punishment for same-sex erotic behavior is neither eminent nor likely. Consequently, same-sex erotic behavior is never extinguished in the natural environment and continues unchanged. In the laboratory, however, the likelihood for punishment suppresses same-sex erotic responses, although it fails to eliminate them. Little evidence exists that aversive sexual conditioning generalizes to the social world. In a host of studies, the absence of any conditioned aversion to same-sex erotic stimuli, no matter which conditioning strategies are employed, suggests that conditioning principles influence the conditions in which behavior occurs but not the sex to which one is attracted. Still, it is possible that an acquired behavior could be highly resistant to change.

One of the more infamous reports of positive results from sexual orientation conversion therapy comes from William Masters and his wife and partner Virginia Johnson (1979), the renowned sex therapists. In their early ground-breaking study, Masters and Johnson investigated the physiological differences in sexual functioning among "gay" men, "lesbians," and "heterosexuals" and described their treatment for sexual conversion. Surprisingly, Masters and Johnson found no differences in sexual functioning between "gays" and "heterosexuals." However, using behavioral modification techniques, they treated a sample of "gay" men and "lesbians" who expressed dissatisfaction with their sexual orientation. The clinicians attempted to neutralize or remove the psychological impediments to other-sex erotic functioning. Therapy required demonstration of *successful sexual functioning with an other-sex partner who entered therapy with the patient*. In the end, Masters and Johnson claimed a greater than 60% success rate for conversion of "gay" men and "lesbians" to heterosexuality. However, in looking at their data more closely, all is not as it appears. Masters and Johnson (1979) noted that at treatment end a significant number of patients made a "full heterosexual commitment" (p. 359), although some patients expressed "sexual ambivalence" (p. 359) and others "returned to their prior homosexual orientation" (p. 401). In recalculating treatment outcome and including treatment dropouts—which Masters and Johnson did not do—as well as treatment failures, the short-term success rate for Masters and Johnson's sexual orientation conversion therapy is more like 48% ($N = 65$).

Other concerns about the Masters and Johnson's sample raise further questions about their conclusions. For instance, the majority of "homosexual" patients who converted to heterosexuality might have more properly been called *bisexual*—Masters and Johnson prefer the term *ambisexual*—because they reported a high frequency of other-sex erotic contacts prior to treatment. In fact, successful conversion to heterosexuality was predicted by a substantial history of pleasurable other-sex erotic behavior, and all treatment participants were required to have an other-sex partner. If other-sex erotic behavior rather than sexual fantasy and attraction is the criterion for successful sexual orientation conversion, then conversion therapy is indeed somewhat successful. Although sexual behavior can be modified, sexual attraction appears resistant to change. Masters and Johnson demonstrated that same-sex erotic behavior can be suppressed, but they did not show that the sex to which one is attracted changed.

The failure of behavioral conversion therapies, like Masters and Johnson's program, provides the strongest evidence that conditioning principles do not account for the development of same-sex erotic attraction. Unfortunately, so-called "ex-gay" organizations and some Christian psychotherapists are not influenced by empirical data disconfirming the efficacy of sexual orientation conversion therapies. Although same-sex attraction is no longer deemed a diagnosable mental illness, a faction of psychotherapists believes that treating same-sex erotic attraction is justified and necessary. One of the most well-known conversion therapists is psychologist Joseph Nicolosi, who operates both the Thomas Aquinas Psychological Clinic and the National Association for Research and Therapy of Homosexuality (NARTH) in California. Nicolosi (1991) coined the term *reparative therapy* to emphasize that treatment corrects a problem that prevents full psychosexual maturity. Impressively, Nicolosi (1991) claims to cure one-third of his "gay" patients—which is substantially less than even Masters and Johnson's recalculated success rate. Another one-third of his patients show significant improvement (whatever that means), and one-third fail to change at all. Despite these grand claims, to date, Nicolosi has given only vague estimates of his success and has offered no reliable or verifiable data for examination. His distinctly antigay and unethical clinical practices in conversion therapy earned him a dismissal from membership in the American Psychological Association. Although Nicolosi is an exception among mental health practitioners, at least one psychologist, University of Georgia professor Henry Adams, asserts that clinicians have an obligation to help "gay" patients try to change their sexual behavior if such a change is in their best interest (Campos, Bernstein, Davison, Adams, & Arias, 1996)—despite the lack of evidence that sexual orientation conversion therapy has any efficacy!

Reparative treatments employed by "ex-gay" organizations are based on religious conviction and the suppression of temptation, with no pretense of being scientific. Such organizations include Exodus International, Homosexuals Anonymous, Evergreen International, Desert Stream, The Christian Coalition for Reconciliation, Living Waters, and Love in Action, to name a few. Treatment usually involves prayer, Bible study, group discussion about the sinfulness of homosexuality, and suppression of same-sex erotic behavior. Because most of these "ex-gay" organizations are religious-based, they are not required to meet state standards for mental health-care professionals, and people who feel abused by these treatments have little legal recourse. People who have failed religious sexual orientation conversion programs refer to themselves as "ex-ex-gays." One "ex-ex-gay," Sylvia Pennington (1989), poignantly describes the painful emotional and spiritual torture she experienced in the "ex-gay" ministry. Despite tremendous personal desire and social pressure to be "heterosexual" Pennington was unable to change her sexual feelings.

Douglas Haldeman (1991) and Timothy Murphy (1992) have published extensive reviews of the efficacy of sexual orientation conversion therapies. Both note that most conversion therapies are directed toward men and conclude that such therapies are motivated by heterosexism and negative attitudes about homosexuality and, as such, are harmful to patients. Doing harm to patients is unethical and unprofessional for health-care providers. Individuals who are distressed by same-sex erotic thoughts and behaviors are particularly vulnerable to emotional manipulation and abuse by antigay therapists. Psychiatrist Richard Isay (1989) claims that a failed attempt to change sexual orientation contributes to further emotional distress, maladaption, and depression among gay men. In fact, concerns about the ethical use of sexual orientation conversion therapies prompted both the American Psychological Association and the American Psychiatric Association to qualify their use. In 1997, the American Psychological Association stated that undertaking conversion therapy must include discussion with the patient about (a) the lack of efficacy data to support the treatment, (b) the possibility that treatment will make the patient feel worse, and (c) the likelihood that social forces in our culture are to blame for the patient's poor self-image and desire to change sexual orientation. The resolution stopped short of condemning the practice of conversion therapy as unethical. Later that year, the American Psychiatric Association (1998) issued a position statement condemning treatments that assume that "homosexuality *per se* is a mental disorder" or that "the patient should change his/her homosexual orientation."

To summarize this section on conditioning, the use of learning-based aversion and conversion therapies provides consistent evidence that sexual behavior can be suppressed and modified but that sexual attraction itself is resistant to change. By their omission of the topic, learning theorists seem to assume that other-sex erotic attraction develops automatically or *naturally* and needs no explanation. On the other hand, learning theorists depict same-sex eroticism as highly reinforcing, easily learned, and resistant to modification (Feldman & MacCulloch, 1971; Greenspoon & Lamal, 1987). Given this view of that same-sex eroticism is highly reinforcing and easily learned, one wonders why there are so few "gay" people in the world?

Most learning theorists who write about sexual attraction have taken a narrow view of the learning environment and failed to discuss how biology and culture affect learning (see Kauth & Kalichman, 1995; McConaghy, 1987). Astoundingly, behaviorists often omit discussion about the powerful social pressures that shape sexuality. In Western culture, children are raised in families that explicitly and implicitly oppose same-sex erotic tendencies, especially for boys. American adults tease young boys and girls about their respective boyfriends or girlfriends before children are even aware of what such a relationship means. Adults *never* encourage children to have a same-sex boyfriend or girlfriend, even as joke. (Ensuring a child's heterosexuality is far too serious to joke about.) Furthermore, Western discourse on sexuality—evident in advertising, literature, music, television, and movies—continuously bombards citizens with the desirability of male-female erotic relationships. Organized social activities—debutante balls, dances, dinner, dating, and marriage—are all structured around male-female pairings, reminding citizens at every turn that other-sex eroticism is the expected and rewarded lifestyle. Social messages about being "heterosexual" are loud, clear, and consistent. At the same time, messages about being "gay" or "bisexual" are usually negative and also quite clear. At best, a few ambiguous or even mildly positive social messages about same-sex eroticism may be heard above the din of "heterosexual" propaganda.

Given our pervasive heterocentric culture, if the sex to whom we are attracted is entirely a product of learning experiences, how could reinforcement be sufficient or sustained long enough to foster the development of even a tiny minority of people with same-sex erotic feelings? Given the obvious rewards and privileges of heterosexuality in Western culture, if conditioning principles alone are responsible for the acquisition of attraction to a sex, how could "gay" men or "lesbians" fail sexual orientation conversion therapy? The answer seems to be that conditioning principles alone are not sufficient to explain how sexual attraction develops.

2.3 SALVAGING PSYCHOSOCIAL THEORIES OF SEXUAL ATTRACTION

Conventional learning models of sexual attraction—whether psychoanalytic, social, or behavioral—present individuals as devoid of biological influences and immune to cultural forces (McConaghy, 1987). A realistic learning model of sexuality should account for how biology through human evolution has shaped how and what we learn and should include the social context in which sexual learning occurs. Ethological and comparative researchers recognize the interaction between biology and learning in non-human species, acknowledging that efficient adult sexual behavior depends on the *timing* of environmental experiences during a *critical period* of development (Harlow & Harlow, 1965; Lorenz, 1965; Whalen, 1991).[4] Is the human species somehow exempt from this interaction? Is learning not a developmental process for humans?

Ethological and comparative researchers also recognize that within a species each sex is predisposed to make particular sexual associations. We know that men and women are predisposed to make particular erotic associations. Men, for example, are more likely than women to develop sexual fetishes, which suggests that males are disposed to making sexual attributions. Women, being more social than men, are more likely to make romantic attributions to stimuli. This is an area that seems ripe for experiments on erotic conditioning. However, given the evidence to date, the sex to which humans are attracted may be predisposed as well and resistant to modification.

However sexual attraction develops, it seems likely that sexual behavioral preferences are learned and that men in particular make sexual attributions to stimuli. Yet there is no evidence that all stimuli are equally susceptible to sexual conditioning. In fact, there is plenty of data to demonstrate that particular objects or physical features are likely to be sexualized and that the sexes differ in their readiness to make sexual associations. Learning theorists have missed an opportunity to investigate how and under what conditions sexual learning occurs. For example, redheads of the appropriate sex may be a tremendous turn-on for some men and some women, whereas blondes or brunettes have no erotic significance. Why is a particular hair color a CS for those individuals? How did it occur? For others, a particular body type, body part, a way of moving, a laugh, color of the eyes, a manner of speaking, or a style of dress has erotic significance. Numerous physical characteristics—muscles or lack of muscles, voluptuous or slender builds, large or medium sized breasts, dark or white or brown skin color, or green or brown eyes—acquire erotic significance for some people but not for others. Why are humans not turned-on by the same physical features? If people conditioned similarly, would we all desire the same kind of sexual partner? Would we all desire blonde-haired partners and reject partners with dark hair? If physical stimuli were equally susceptible to sexual conditioning, would we experience a sexual response to all women or to all men?

It seems likely that erotic conditioning is selective and varies by sex and per individual. Over time, single erotic attributions are integrated into a *type* of man or woman to whom the individual experiences a sexual response. Age-old evolutionary demands have favored males who make sexual attributions to sex-typed stimuli and favored females who make romantic/affectionate attributions. This does not mean, of course, that women never make sexual associations or that men never make romantic associations.

The available evidence indicates that we do not yet know the extent to which learning influences human sexuality. Learning theorists in the past were misdirected by negative cultural beliefs about homosexuality and attempted to reduce all of sexuality to a conditioned response. Despite their grand explanation, previous learning theorists who wrote about sexual attraction largely ignored how social context influences learning and focused exclusively on the sex of attraction, although the sex of attraction may be the least consequential element of sexual learning. The peculiar social priorities of Western culture in this century misled scientists, physicians, and researchers into focusing attention on the "mote" in the eye of human sexuality and ignoring the more obvious and interesting "beam."

Before describing a new model of sexual attraction that integrates biologic and environmental influences and learning experiences, we must review a social theory of sexual attraction that has dominated theoretical efforts in the field of sexuality for the past 20 years: *social constructionism*. This is the topic to which we now turn.

CHAPTER 3: CULTURE'S CHILD: A SOCIAL CONSTRUCTIONIST MODEL

The social world contains many realities that do not exist in nature. — David Halperin, *Saint Foucault*

People invent categories in order to feel safe. White people invented black people to give white people identity. . . . Straight cats invent faggots so they can sleep with them without becoming faggots themselves. — James Baldwin, *A Dialogue*

There is no such thing as a homosexual or a heterosexual person. There are only homo- or heterosexual acts. Most people are a mixture of impulses if not practices, and what anyone does with a willing partner is of no social or cosmic significance. — Gore Vidal

The predominant social model of sexual attraction today is a collection of theories called *social constructionism*. To fully describe the model's various and complex tenets and implications, I begin by examining a radical version of the theory in the first section of this chapter. In the later part of this chapter and again in Chapter 5, I describe a less ideological, *middle ground* social constructionism (Vance, 1991).

Social constructionism is not a theory about sexuality; it is an epistemological position about the nature of knowledge. Social constructionism has developed from a veritable garden of diverse disciplines, including anthropology and sociology through symbolic interactionism, political science through Marxist history, social psychology and sociology again through labeling theory, and the postmodern humanities in the form of structuralism and poststructuralism, feminist and gender theory, and gay/queer studies (Burr, 1995; Vance, 1991). Most recently, social constructionism, or *deconstructionism*, has been a favored tool of social activists, including feminists and gay or queer writers, who use constructionist rhetoric to deflate an opponent's position and promote their political agenda. As a tool for persuasion, constructionism is fairly successful.

There is no one theory of social constructionism as practiced; it is a collection of ideas that share common assumptions. Rarely, are the assumptions underlying the theory ever discussed. However, identifying and understanding these assumptions are central to grasping the limitations of social constructionism in explaining sexual attraction. For this reason, I list the underlying assumptions of social constructionism now in Table 1 and will point them out again throughout this chapter. Foreknowledge of these assumptions will help the reader disentangle a number of conceptual subtleties and effectively evaluate the theory's general claims.

Principally, social constructionism asserts that subjective meaning is the *only* type of knowledge that we truly possess (Berger & Luckmann, 1966; Foucault, 1978/1990). Objectivity is a social contrivance—a conventional way of speaking about particular experiences. Concepts and categories are socially agreed upon "facts" and they do not explain social reality so much as *create* it. Broad concepts and values like democracy, individualism, civil rights, medicine, marriage, science, and sexuality may become institutionalized, as is the case in Western culture, but their general acceptance does not make these concepts universal or *natural*, although they seem that way to us.

As noted in chapter 1, *culture* broadly encompasses ideas about social reality, including customs; mores; cherished values about, for example, individual autonomy and civil rights; social institutions like marriage and democratic government; social role; gender role; ethnicity; class; and, most

TABLE 1
Assumptions of Social Constructionism

1. *People are constructed entirely from their continuous interactions with the social world.*
2. *People are generally passive respondents in the social discourse.*
3. *Sex is largely recreational.*
4. *Social labeling of sexual behavior and internalization of the "homosexual" role creates "gay" men and "lesbians."*
5. *Biology sets potentialities or preconditions on sexual behavior, although the effect of these limits is small and fixed.*
6. *People possess diffuse undirected sexual feelings that are channeled and developed through social discourse in a given culture.*
7. *Sexual relationships can be structured an infinite number of ways.*
8. *Labeling theory of same-sex eroticism is a causal narrative.*
9. *Social constructionism is a critical theory.*

recently, sexual identity. Culture also includes the conveyance of ideas and information through various means. Culture is communicated visually through the arts and by word of mouth via folklore, storytelling, and religious practice. Culture is communicated through the written word in books and literature and in technological developments. Culture is at heart a social experience, like class or gender; however, culture is also transmitted through general social experiences like war, poverty, political domination, slavery, famine, plague, and, beginning in the 20th century, mass media, including radio, television, and now the Internet. Different cultures produce different *social realities*. Social reality is more than acculturation and socialization; it includes how people think and feel, how they organize phenomena, what meaning events have, the place of individuals, and what possibilities are available to individuals given their social role. Constructed social reality is not like an article of clothing that can be swapped easily with another to create a new fashion. Social reality is the whole of life; it is what we know and accept to be true. Through numerous interactions with others and constant contact with the culture, particular roles and personal identities are created for each of us (Berger & Luckmann, 1966; Foucault, 1978/1990).

Although not exactly a theory of sexuality, recently constructionist writers have focused on the historical and political development of categories of sexual classification, including *sex*, *gender*, and *sexual orientation*. Again, a major appeal of social constructionism is its political utility in challenging—dramatically and effectively—what we believe to be true. Consequently, social constructionism has been adopted by academic and social activists such as feminists, gay rights activists, and AIDS activists who desire broad social changes in Western culture. More specific to our purpose, however, social constructionism has been the reigning social theory of sexuality for over 20 years.

Although less intuitive and strange to some, social constructionism is grounded in discernible assumptions about knowledge, experience, and individuals. These assumptions, as listed in Table 1, should become apparent in the subsequent discussion. The form of this discussion is a pseudo-debate between social constructionists and so-called (biomedical) essentialists. This mock debate is a favorite rhetorical device for social constructionists to present radically simplistic versions of their own ideas in contrast to a stereotypic biomedical position. Only social constructionists participate in this debate. Although the format is slightly awkward, I choose to simulate the debate to further illustrate how constructionists utilize the theory as a means of persuasion.

3.1 THE ESSENTIALIST-CONSTRUCTIONIST "DEBATE": PRINCIPLES OF SOCIAL CONSTRUCTIONISM

Most social constructionists who write about sexuality employ the rhetorical debate (see Epstein, 1987; Weinrich, 1987b). This debate is about the nature of knowledge. A great amount of constructionist literature rails against biomedical explanations of sexuality, called *essentialism*. Some writers use the terms *realism* and *nominalism* rather than essentialism and constructionism (Boswell, 1982-1983; Hacking, 1986; Weinrich, 1990). The semantic distinctions between these word-pairs are small and trivial to the direction of this text. However, it is important to note that *realist* and *essentialist* are pejorative terms in the constructionists' mock debate (Boswell, 1982-1983). Biomedical writers do *not* use these terms to refer to themselves.

So-called *essentialists*—evolutionists, biologists, and many social scientists—do not participate in this debate; constructionists argue both sides themselves. However, through this rhetorical device, the principles of social constructionism emerge, in contrast to the biomedical essentialist position. I will start with the essentialist position.

According to social constructionists, biomedical essentialists are defined by two beliefs: *the world operates in an orderly fashion according to universal laws of nature*, and *all objects can be grouped into naturally occurring categories* (Dupré, 1990; Plummer, 1981; Wieringa, 1988). For biomedical essentialists, understanding the world and its inhabitants is a function of discovering the basic underlying elements and order of each phenomenon (Dupré, 1990; Epstein, 1987). To understand a phenomenon by reducing it to its basic components is called *reductionism*. Biologists and neuroscientists, for example, reduce complex human behavior to chemical processes between cells and between groups of cells. Through dispassionate systematic observation, essentialists are said to believe that knowledge and objective Truth are revealed. Because the world is assumed to be orderly, essentialists can generalize from controlled observations of phenomena within our reach to phenomena that cannot be directly observed, such as internal processes, microscopic organisms, celestial bodies, or even ancient societies.

More to the topic at hand, social constructionists assert that biomedical essentialists view sexual attraction as an inherent universal biologic force that exists in all animals, including humans (Epstein, 1987). That some people experience sexual attraction only to males, some only to females, and still others to both sexes, denotes genuine, differences among people. Essentialists are said believe that different sexual orientations are a function of particular genes, hormones, or other biologic events. Thus, labels such as "homosexual," "heterosexual," or "bisexual" classify different *natural kinds* of people, not just variations in sexual desire. Essentialists are also said to believe that although different erotic types of people have always existed, past cultures regarded these distinctions as unimportant and did not bother to name them.

By contrast, social constructionists view sexual categories such as "heterosexual" and "homosexual" as historically specific, culturally endorsed roles or proscriptions to regulate identity and social conduct (McIntosh, 1968/1990; Foucault, 1978/1990). Sexual desire itself may even be socially constructed. However, if sex drive is rooted in human biology, there are no essential or biological characteristics that make "homosexual" people different from "heterosexual" people. Social constructionists point out that these sexual orientation categories originated in late 19th and early 20th centuries in Western culture. Constructionists argue that society, not biology, force people to fit into prefabricated sexual orientation categories to meet cultural needs. Supposed distinctions between people of different sexual orientations simply reflect the current common beliefs that individuals differ in the direction of their sexual feelings, as well as differ across various sex-typed characteristics, and that sexual orientation is inherent and immutable. However, as constructionists note, cultures that give little importance to the sex to which one is attracted structure sexual relationships in different ways. In those cultures, attraction is generally not thought to be exclusive to one sex, and people are not identified by the sex to which they are attracted. Labeling sexual attraction as *sexual identity* is a contemporary Western idea that carries with it many assumptions that are specific to this era and culture. Therefore, talking about "gay" or "heterosexual" people in other cultures is both inappropriate and misleading.

Social constructionists usually confine their discussion to a biomedical version of essentialism and to the study of homosexuality. A rare exception is social historian Jonathan Katz (1995) who devoted an entire book to the social construction of heterosexuality. However, for the most part, social constructionists contrast their ideas with archaic biomedical and psychiatric theories of male homosexuality. These early essentialist theories are portrayed as rigid, dogmatic, empirically careless, and politically oppressive, which they are (Foucault, 1978/1990; Greenberg, 1988; Katz, 1995; McIntosh, 1968/1990; Padgug, 1979/1990; Plummer, 1981; Weeks, 1985). Constructionists rightly conclude that old biomedical theories of homosexuality are static and deterministic products of a particular historical period. Contemporary interactive theories of sexuality are never mentioned by social constructionists (Wieringa, 1988). Instead, social constructionists describe how psychoanalysts and medical practitioners created the diagnostic label *homosexual* to pathologize same-sex erotic attraction, although neither psychiatry nor medicine currently hold these views (Bayer, 1987). Although psychiatry and medicine contributed greatly to the birth of the "homosexual," it is arguable that these institutions are still the primary supporters of this concept today. Century-old ideas about sexuality, of course, appear naïve and unsophisticated (Wieringa, 1988). In short, social constructionists created the biomedical essentialist as a straw man in order to make their position appear more rational and convincing (Dynes, 1990).

Essentialism and constructionism are not complementary concepts, and the debate was not meant to be one. Rather, arguments in favor of social constructionism are designed not to inform but to persuade. Indeed, social constructionists assert that acceptance of an idea as valid and legitimate is not the result of strong empirical data but is due to the effect of convincing persuasion (Gergen, 1985). Data possess no inherent Truth; Truth, according to the social constructionist theory, is what you believe is true.

From this brief overview of social constructionism, I now turn to several key principles of constructionism that derive from the debate. The strengths and weaknesses of these principles and their relationship to sexual attraction will be discussed.

3.1.1 Categories and Concepts Are Not Universal

So-called biomedical essentialists are said to believe that things and people fit *naturally* into universal and objective categories—for example, the sexual categories "homosexual" and "heterosexual." Although other cultures use different terms or fail to remark on distinctions between sexual orientation, social constructionists describe essentialists as accepting that differences in sexual attraction characterize *natural* kinds of people. In other words, the biomedical essentialist position is said to be that "homosexuals" and "heterosexuals" exist even in cultures that fail to recognize those concepts or distinctions.

By contrast, social constructionists note that same-sex erotic behavior is relatively common in many societies. Yet almost no other culture employs sexual categories comparable to "homosexual" and "heterosexual" (Padgug, 1979/1990; Plummer, 1981). And why would they? Social constructionists argue that sexual orientation categories like "homosexual" and "heterosexual" do not reflect internal traits but signify the unique and peculiar way that contemporary Western culture views sexuality. These categories embody particularly recent Western values—individualism, capitalism, egalitarianism, social purity, normative sexual behavior, and affectionate marriage—that did not exist in this form prior to the late 19th century (Marshall, 1981; Trumbach, 1988; Weeks, 1981). The fact that numerous cultures organize sexual relationships and define sexual acts in so many different ways argues against universal categories of sexual attraction, according to constructionists (Padgug, 1979/1990). According to this argument, traditional Navajo male *two-spirited people* who cross-dress and have sexual relations with men have little in common with the drag queens who populate Greenwich Village. Ancient Athenian men who were married and had boylovers are quite unlike married men from the suburbs in the late 20th century or early 21st centuries who coach Little League baseball on Saturdays and download child pornography from the Internet in their spare time. Likewise, Sambian men in New Guinea, who have intercourse with their wives and are fellated by young boys are not bisexuals in the Western sense. Applying terms like "gay" or "heterosexual" or "bisexual" to people in pre-industrial non-Western cultures is anachronistic and misleading because the ideas embodied by contemporary sexual identity terms are alien to those cultures.

3.1.1.1 The Context

Social constructionists rightly point out that the meaning of sexual behavior depends upon its context (Gergen, 1985); this rule is no less true when the behavior occurs in ancient cultures. There are no intrinsic sexual behaviors; rather social parameters determine when an event is *sexual*. Parameters that define a *sexual* event include the participants' gender, age, class, culture, and relationship to each other, as well as the particular act involved (Padgug, 1979/1990). The social context defines what is *sexual* and what is not. Constructionist John Gagnon views each cultural context as completely independent of others, with only superficial similarities, and believes that the sexuality that develops within each culture is equally distinct. Gagnon (1990, p. 182) states,

> The (sexual) conduct could not have been learned in the same contexts, it is not practiced for the same purposes, it is not maintained by the same social forces, and it does not cease to be practiced at the same moments in the life course for the same reasons. The social construction of sexuality (and even if sexuality exists as a separable domain in a culture) and its connections to nonsexual conduct are specific to the cultural and historical circumstances of a particular social order.

According to Gagnon, sexual behavior is not similar across cultures. What is more, to apply contemporary theories of sexuality to cultures other than our own constitutes an "act of theoretical hubris" (Gagnon, 1990, p. 182).

The point here is that the same behavioral act may be defined as *sexual* in one context and *non-sexual* in others. This is as true within a culture as it is across cultures. Kissing, hugging, or body rubbing are deemed sexual acts under some circumstances and for some individuals but not for all. Oral stimulation of a penis may appear to be a universal sexual act, but recently this perception was called into question in our own culture. In 1998, President William Jefferson Clinton testified under oath before a grand jury that he did not have a *sexual* relationship with former-White House intern Monica Lewinsky (Starr, 1998). Lewinsky, however, testified earlier that during their 18-month affair she performed oral sex on Clinton on several occasions. Lewinsky noted that Clinton never reciprocated. She also described several episodes of masturbatory sex while she and the president talked on the telephone. She denied that they ever engaged in vaginal intercourse. What is interesting here is that both Lewinsky and Clinton *denied* that oral sex constituted *sex*. Both held a legalist definition of a sexual relationship as involving only vaginal intercourse; since they had not engaged in vaginal intercourse, they had *not* had sex by this definition.

A few months later, much to the surprise of many adult Americans, researchers reported that university students also have a relative definition of sex similar to Lewinsky and Clinton. Among a random stratified sample of 599 students designed to represent the undergraduate population at a state university in the Midwest, 60% indicated that oral-genital contact did *not* count as having "had sex" with a partner (Sanders & Reinisch, 1999). What is more, 19% of students believed that penile-anal intercourse did not constitute sex. Study researchers Stephanie Sanders and June Reinisch pointed out in their conclusion that Americans do not agree on what constitutes sexual activity and the relational context of the behavior defines its meaning as *sexual*, consistent with social constructionist principles. Significant relational variables, at least for university students, include whether the behavior occurred in an established relationship, the relevance of orgasm to intimacy, issues of consent, cohort and socioeconomic status of participants, and the potential costs and benefits of identifying a behavior as sex. In other words, *sexual activity* can be difficult to recognize, even if you "see it." Therefore, one must be cautious when attempting to define someone else's sexual behavior.

The risk of misinterpreting behavior as *sexual* increases when observing cultures outside of one's native culture. Other cultures may also define very intimate physical acts as non-sexual under certain circumstances. As discussed before, Sambian men describe the act of boys fellating men as "growing men." Boy-insemination is viewed as masculine, anti-female, and social, although affectionate (Herdt, 1981, 1987). According to informants, male-male fellatio during boy-insemination rituals is only peripherally sexual, although, as we discussed in chapter 2, there is some evidence that these rituals are more sexual than informants admit. Still, do we disregard the Sambians' account of their behavior? Do we doubt their honesty? Do we reinterpret their meaning? When anthropologist Gilbert Herdt (1981) initially described boy-insemination rituals among the Sambia, his cultural biases slipped through in his portrayal of the rites as *homosexual* rather than describing them as same-sex. In doing so, he implied, carelessly, that these rituals are explicitly sexual and are performed by males who are attracted to other males. The rituals were not "homosexual" ones for the Sambians. In later writings, Herdt acknowledged his conceptual and linguistic error and referred to ritualized male fellatio among the Sambia as *boy-inseminating rites* (1984/1993a,b, 1991; Herdt & Stoller, 1989).

The lesson is that the words we use to describe another's behavior may more accurately illustrate what we *think* about the behavior. Rather than impose a definition, we should look for a behavior's meaning from informants and the social context, although these accounts may require some interpretation.

If there are no universal sexual orientation categories to account for erotic feelings and sexual attraction, what kind of sexual "stuff" is left? Social constructionists Kenneth Plummer (1981) and Michel Foucault (1978/1990) both suggest that human sexual attraction is general and undirected. However, powerful and constant social forces in everyday life channel, truncate, restrict, nourish, and sexualize our various experiences. Over time, individuals learn to attend to particular *sexual* desires and ignore or suppress other desires that are culturally discouraged or even prohibited. Thus, human sexual attraction is initially a set of diffuse desires that are molded and defined by cultural values; some desires are experienced as *sexual* and appropriate under certain circumstances, while other desires are experienced as non-sexual or as erotic but taboo. Once established and supported by the culture, constructed categories such as *man*, *woman*, *husband*, *wife*, *heterosexual*, *homosexual*, and *black* live as "real fictions" (Foucault, 1978/1990; Weeks, 1988). Cultures train their members to fill the social roles and identities that are available. However, roles are not enduring or unalterable, although they may span several generations. According to constructionist thought, social identities are "real" in the sense that they are treated as valid and true by members of the culture. Some identities such as *gay*, *lesbian*, *woman*, and *black* are stigmatized and carry painful consequences for those so labeled. In contemporary Western culture, "heterosexuals" benefit from explicit privilege and honor through public acceptance and institutionalization, whereas "gay" men and "lesbians" are often ignored, victimized, and silenced.

Sex researchers Philip Blumstein and Pepper Schwartz (1990) hold a similar view of human sexuality. Based on data from two large studies in the 1970s and 1980s of "bisexuals" and "gay" and "straight" couples, Blumstein and Schwartz conclude that sexual attraction is fluid and not fixed. Their first study of 150 men and women investigated antecedents of sexual identity and bisexuality (Blumstein & Schwartz, 1976a,b). The second large study collected data from nearly 1,000 "gay" couples, 800 "lesbian" couples, 3,600 "heterosexual" married couples, and 650 "heterosexual" cohabitators (Blumstein & Schwartz, 1983). In these two studies, many "gay" men and most "lesbians" had or continue to have other-sex partners during their lifetime. What is more, some "heterosexual" men and a few "heterosexual" women reported having same-sex erotic partners despite an exclusive sexual orientation. In Blumstein and Schwartz's view, sexual identity labels limit full sexual desire—which would seem to be bisexual—and prohibit romantic relationships that run counter to sexual identity. They believe that once people grow accustomed to their sexual identity, their erotic feelings and sexual behaviors become more consistent with that label.

In short, the social constructionist position is that sexuality is culturally defined and unique—each real, but none essential; there are as many sexualities as there are cultures. Sexualities are like sand on the beach or as variable as snowflakes. To illustrate this point, social constructionists describe various cultures that define sexuality differently than contemporary industrial Western culture (Padgug, 1979/1990; Plummer, 1981; Marshall, 1981; Trumbach, 1988; Weeks, 1981).

The problem is that variation in sexuality across cultures does *not* refute the idea that sexual attraction has an essential biological core (Stein, 1990). Biomedical theorists of sexual attraction do not need to explain sexual diversity across cultures; they need only to allow enough flexibility in their theories to account for cultural variations in sexual behavior. Sexual tastes and behavioral acts could vary across cultures, even if sexual attraction is strongly biological. Sex drive, for example, could be a biological process, and a few social constructionists even concede this point (Plumber, 1981; Weeks, 1988), although at least one (Gagnon, 1990) claims that the biological processes involved would themselves vary across cultures. If we assume for the moment that sexual drive is biologically based, all sorts of social and behavioral orchestration could develop around it. The lack of similarity in the social manifestations of sex across cultures does not contradict the supposition that a biological sex drive underlies sexual behavior. Likewise, the fact that contemporary Western cultures identify people by which sex they find sexually attractive, while most pre-industrial non-Western cultures define people and sexuality in other ways, by no means denies that biological processes influence the direction of sexual attraction. Contrary to social constructionist claims about biomedical essentialist theory, a biologically based sexual attraction does *not* prohibit variation in the social expression of sexual behavior.

3.1.1.2 Exceptions or New Rules

Historian John Boswell (1982-1983) takes issue with social constructionist beliefs that *variation in conceptual categories means that they are not grounded in recognizable "real" distinctions* and that *the absence of a name for a phenomenon suggests that distinctions are unrecognized in that culture.* Such beliefs fail to reflect the social reality of many cultures, he claims. Boswell first notes that well-established concepts, such as *Jews*, *blacks*, the *blind*, and *family*, have varied considerably across cultures and over time. The term *Jews*, for example, has referred to a host of odd and contradictory characteristics over the centuries without diminished use of the concept. In addition, despite various cultural configurations of *family*, we easily recognize families in diverse cultures. Boswell asserts that *sexual orientation*, like *family*, is a useful concept that denotes practical distinctions between people.

Second, contrary to social constructionist claims, Boswell insists that *exclusive* same-sex erotic attraction existed in a variety of ancient cultures and was recognized as such. Boswell (1980, 1982-1983, 1990) cites several ancient references to exclusive same-sex erotic attraction among individuals, and couples, and erotic kinds of people that were understood by the culture. For instance, in the ancient world and in pre-industrial Western cultures, it was not unusual for same-sex partners to form unions similar to marriage, complete with ceremony, exchange of vows of love and fidelity, and church approval under some circumstances (Boswell, 1994; Brooten, 1996; Leupp, 1995). Boswell (1994) notes that same-sex unions were performed in Roman Catholic churches well into the 13th century. While evidence of church-sanctioned same-sex unions is rare after the 13th century, moralist Michel de Montaigne (1533-1592) reported the occurrence of such a ceremony involving two dozen male couples at the Church of St. John in Rome in 1578 (Boswell, 1994). What is more, Emperor Nero twice married a man in public ceremonies, and by the later part of the first century C.E., Juvenal regarded male unions as commonplace in the Roman world (Boswell, 1994). Unlike marriage between a man and a woman, which was a legal arrangement of property and heirs, same-sex couples in the ancient and premodern world did not become each other's dependent or property. Because of these distinctions, Boswell hypothesizes that same-sex unions were more likely than marriages to be motivated by affection, because economic or property inheritance and paternity of children were not involved. Neither marriages nor same-sex unions were necessarily exclusive sexual relationships and did not signify exclusive sexual attraction. Unlike marriage, same-sex unions were not the norm in the ancient and premodern world, and their existence suggests that for some people marriage could not or did not meet their needs. Furthermore, the existence of same-sex unions in pre-industrial cultures in which marriage was the norm also suggests that participants experienced an affectionate preference that was not requisite for marriage.

Classicist Eva Cantarella (1992) disagrees with Boswell about how common male unions were in the ancient world. She argues that male-male unions, at least, were rare and largely an activity of privileged citizens. On the other hand, Bernadette Brooten (1996), who has extensively researched same-sex love between women in the early Christian world, like Boswell, concludes that same-sex unions in general were not unusual. Both Christian writer Clement of Alexandria and the second-century astrologer Ptolemy of Alexandria, for example, noted that same-sex female unions were common practice during their lifetimes (Brooten, 1996). Brooten further notes that same-sex love by both sexes is documented in the texts of erotic spells, which were wildly popular in the ancient world. Thus, it seems clear that, contrary to social constructionist claims, ancient and premodern Western cultures recognized that some people had largely same-sex erotic interests.

In addition, there is evidence that exclusive same-sex attraction and same-sex unions were recognized in premodern non-Western cultures. Through the Tokugawa period (1603-1868) in Japan, male-male love is well documented (Leupp, 1995). Similar to the ancient Greek custom of *paiderastia*, the love of boys among men was practiced in Japan, beginning around 806, and known as *nanshoku* (boy-love), *wakashudo* or *jakudo* or *shudo* (the way of youths), *nando* (the way of men), *bido* (the beautiful way), and *hido* (the secret way). In contrast to *paiderastia*, *nanshoku*, as the practice was more commonly known, was more explicitly a sexual relationship. Furthermore, *nanshoku* relationships were likely to continue after the younger partner reached adulthood, unlike *paiderastic* relationships, which usually ended, when the younger partner acquired secondary sex characteristics. Some Greek paiderastic relationships, however, are documented between adult men.

By the 17th century, legal male unions were commonplace in Japan, largely due to the popularization of *nanshoku* by samurai. *Nanshoku* was so closely associated with samurai warriors that it became known as "the past time of the samurai" (Leupp, 1995). Male partners in a *nanshoku* relationship formed a "brotherhood bond," in which the "older brother" and "younger brother" swore oaths of loyalty and signed written vows during a formal ceremony. "Brotherhood" contracts sometimes specified a restriction on the *number* or *sex* of future sexual partners, if any partners were allowed at all, for one or both men. A contractual ban on future male sexual partners suggests that emotional infidelity was only at risk with another male, possibly because sexual relationships with females did not count as a violation of the vows, or that only males were potential sexual partners. The emphasis on romantic love and courtship within *nanshoku* relationships differed substantially from marriage. Not until the late 17th century did male-female courtship and marriage in Japan begin to mimic the romance and passion of *nanshoku* relationships. Although all men in Tokugawa Japan were expected to marry women, most maintained concurrent sexual relationships with boys and other women. However, according to Leupp (1995), a minority of Japanese men maintained *exclusive* erotic relationships with men, and some *nanshoku* relationships were lifelong.

Women in Tokugawa Japan could also be samurai and could enjoy sexual activity with women at female brothels. Even so, Leupp (1995) notes that there is little evidence of female-female marriage during this period of time. Still, male unions indicate that same-sex romantic attraction and a preference by some for male sexual partners was recognized by the culture, contrary to constructionist beliefs.

Boswell (1980, 1982-1983, 1990) asserts that the ancient Western world recognized distinctions in sexual attraction that are not so different from our own. Evidence of this, according to Boswell, can be found in Plato's (427-347 B.C.E.) *Symposium*. Plato has Aristophanes describe a creation myth in which the gods split apart *three types* of early dual humanoid creatures into *half*-humans. Each half-creature longs to be reunited with its spiritual and emotional complement, which was understood as the basis for romantic attraction. For some people, their half is someone of the other sex; for others, their complement is someone of the same sex. Thus, in this creation story, Plato suggests that people are essentially other-sex oriented or same-sex oriented. Aristophanes goes on to say that some men are "inclined to be a boylover or a beloved, as he always welcomes what is akin" (Warmington & Rouse, 1984, p. 88).

Plato is one of the few ancient writers to develop a comprehensive distinction between erotic types of people. For him, common love (by men) was the love of women, while heavenly or ideal love was the love of boys. At this time in Greece, the practice of *paiderastia*, or the love of boys by men, had become more sexual and even exploitative. Plato's *Symposium* in many respects was a treatise for ideal male-male love and a return to traditional values. In the *Symposium*, Pausanias explains that if the *erastēs* (the older lover) is only in love with the body he will lose interest when the *erōmenos* (the younger beloved) reaches adulthood. Ancient Greek custom dictated that a paiderastic relationship ended when a boy reached puberty. Yet, Pausanias notes that if the *erastēs* loves the soul or good character of the *erōmenos*, his love can last a lifetime.[1] Pausanias had good reason to say this; at the time, he was *erastēs* to Agathon, a grown man. This apparently was not uncommon; other Greek men continued their paiderastic relationship after their beloved became a man, and some men appeared to prefer the love of men over women. Zeno and Parmenides continued their youthful erotic relationship as adults, as did Agathon and Euripides, and Crates and Polemo. Alexander the Great (355-323 B.C.E.), Stoic philosopher Zeno (late fourth century to early third century B.C.E.), and philosopher Bion (third century B.C.E.) preferred male lovers, as did Plato (Greenberg, 1988; Halperin, 1990).

Historian Kenneth Dover (1978/1989) claims that the ancient Greeks understood that men possessed characteristics in degrees and, therefore, a man could be more inclined to boylove than to love of women. Xenophon used the term *tropos*—meaning character, disposition, or inclination—to refer to the enthusiastic boylover Episthenes (*Anabasis*, 7.4.7; Dover, 1978/1989, p. 51). However, most men in the ancient Western world had relationships with both males and females, even in cases where a same-sex erotic relationship was the primary one. For example, Aristogeiton and his lover Harmodius, the assassins of Hipparchus, the tyrant of Athens, both had female mistresses (Cantarella, 1992; Dover, 1978/1989; Halperin, 1990). Athenian politician Timarkhos was charged by a foe in 346 B.C.E. with being a male prostitute because he had taken the receptive sexual role with other men, although he had had many female sexual partners. Alcibiades, the Athenian general and politician who harbored an unconsummated love for Socrates, had sexual relationships with both women and boys. Socrates, himself a married man, consistent with social custom, patronized female prostitutes and loved boys.

Several ancient astrologers referred to people who appeared to have exclusive same-sex erotic tastes and explained their inclinations as combinations of star signs (Brooten, 1996). Writers in ancient Rome and premodern Arabia recognized that some men were exclusively attracted to one sex—either males or females—although most men were attracted to both sexes (Boswell, 1982-1983, 1990). In a 9th-century text, Qusta ibn Luqă compared people on 20 psychological characteristics, including sexual object choice. He noted that some men are "disposed towards" (*yamû lu ilă*) women, some toward men, and others toward both. Saadia Gaon, a Jew living in 10th century Muslim society, who was familiar with Plato's works, wrote about the desirability of "passionate love" but seemed to refer only to male-male relationships (Boswell, 1982-1983). Gaon described the love that men have for their wives as good, but not passionate; male-male love, on the other hand, was passionate. Thomas Aquinas believed, as did Aristotle, that some men were congenitally inclined to love boys. And in 12th century Europe, Allain of Lille, commenting on the sexuality of his peers, said, "Of those men who employ the grammar of Venus there are some who embrace the masculine, others who embrace the feminine, and some who embrace both." (Boswell, 1982-1983, p. 105).

In addition, several ancient literary texts present fictional characters who have exclusive same-sex erotic interests. Medieval novelist Longus described a character in *Daphnis and Chloe* whose same-sex erotic passion was "by nature" (*fusei*) (Boswell, 1990, p. 23). Up through the Renaissance, the term *Ganymede*, a reference to the erotic relationship between the shepherd boy and Zeus, generally referred to exclusive male-love. In the popular poem *Ganymede and Helen* from the High Middle Ages, these two mythological figures debate the merits of male-male love versus male-female love. Ganymede advocates for what appears to be exclusive male-male love. Fictionalized debates about the merits of male-male love versus male-female lover were quite popular in the premodern era. Again, contrary to the assertion of social constructionists that cultures outside of the contemporary West did not have concepts, categories, or words for erotic *kinds* of people who were sexually attracted to one sex, this is not the case.

As a brief but significant side note, Boswell (1990) comments that *exclusive* other-sex erotic attraction is much more difficult to identify than *exclusive* same-sex attraction through historical records. It is curious that the lack of historical evidence regarding exclusive other-sex erotic attraction has not raised questions about whether "heterosexuals" existed in the ancient past. Boswell asserts that historians and theorists require a higher standard of evidence for confirming same-sex erotic attraction than for confirming other-sex attraction. If treated comparably, the rarity of ancient documents that identify exclusive male-female sexual attraction might lead observers to conclude that exclusive other-sex attraction was uncommon or just uninteresting. Yet the infrequent mention of exclusive other-sex erotic attraction in history has *never* sparked a debate about the biological foundation or essential existence of male-female attraction. The controversy is specific to exclusive same-sex attraction.

In an extension of the argument that sexual attraction was recognized and understood in the past without being labeled, Boswell (1982-1983) claims that ancient Romans were aware of the physical property that later was called *gravity*, long before its discovery by Newton. Boswell believes that contemporary concepts have a long prior history as loose ideas that are either unnamed or have various names. Similarly, he believes that the absence of terms in other cultures for exclusive same-sex or other-sex erotic attraction does not indicate that such conceptions are unrecognized and nonexistent. Rather, the weak linguistic emphasis on the direction of sexual attraction in pre-industrial non-Western cultures may reflect cultural indifference or the commonness of the phenomenon. Although language calls attention to particular sexual behaviors, Boswell argues that a lack of explicit terminology is not sufficient to conclude that distinctions between sexual attractions were irrelevant.

3.1.1.3 Cultural Similarities

Although they argue for dissimilarity across cultures in how sexual attraction develops and is conceived, social constructionists are unable to account for cultural similarities in erotic lifestyles, especially among people with same-sex erotic interests (Wieringa, 1988). For example, a common form of erotic relationships in the past involved a man and a boy. Wayne Dynes (1990), a critic of social constructionism, believes that the persistent pederastic model of boylove throughout history and across cultures may, in fact, reflect a primary same-sex erotic attraction among some men. Boswell (1982-1983) even asserts that ancient pederastic models of male-male love were not so different from the relationships of contemporary "gay" men. He notes that the Greek ideal of a generational difference between males in paiderastic relationships was often not met and the actual age difference between partners was small. Furthermore, although the love of a boy was supposed to stop at puberty, there are many examples from the ancient world in which men continued their relationships for many years. Men and women who lived a relatively exclusive same-sex lifestyle in past cultures differed from the norm and probably endured social pressure to be more conventional. Ancient Greeks and Romans who violated the social norm of "bisexuality" by restricting their sexual encounters to one sex, whether to males or to females, experienced gentle ribbing, ridicule, and even open condemnation for their behavior (Cantarella, 1992; Greenberg, 1988).

Saskia Wieringa (1988) and Rictor Norton (1997) point out another common erotic type across cultures—the *mannish* or *butch* woman and the *effeminate* man. They both claim that the *mannish* woman and *effeminate* man have deep historical roots and are not just particular social types of people. The *mannish* woman and *effeminate* man have a "nature" and sexuality that sets them apart from others. Usually these stereotypes referred to women who assert themselves socially or dress like men and to men who play the "female" (receptive) role in anal intercourse. Wieringa (1988) attributes the continuity of this sexual type of person to the number of cultures that polarize gender roles. Norton (1997) extends these transcultural sexual types a step further by collectively referring to them as "queers." By "queer," he means people who know that their primary same-sex erotic attraction differs from the norm, however it is defined, and who do not fit conventional gender roles for their sex, setting them apart from others. For Norton, "queerness" is innate, but its expression as an identity, a third-gender or as mollies, is contextual. "Queerness" then is a sensibility of personal difference and the common associated behaviors that reflect those differences. Norton claims that rigid gender roles and social marginalization of those who have "queer" desires foster similar ways of being.

Although history provides many examples of sexual types like the *mannish* woman and *effeminate* man who violated conventional gender roles, less attention has been paid to non-normative sexual types who honored conventional gender roles. Contrary to social constructionist assertions, several terms have been used to refer to the masculine partner of same-sex relationships, including *erastēs, paedico, pederast, ingler, machista, older brother, lover,* and *top* (Norton, 1997). Yet masculine men who love men are less distinctive throughout history because their social behavior did not set them apart from others. Members of the Sacred Band of Thebes, the elite army of 300 male lovers, were idolized for their heroism in battle and seen as ultra-masculine (Boswell, 1980). Apparently, the Theban recruiters had little trouble identifying the *type* of man that they wanted in their army. It was no accident that *all* of the men were in an erotic relationship with a fellow soldier. But, aside from being noble and brave, we know little about the characteristics of these masculine men-loving men. Emperor Hadrian, a man's man, was not disparaged for his obsessive affection of his Greek lover Antinous, even as he populated the empire with Antinous's likeness and made him a deity. The brave Christian knight, Richard the Lion Hearted, a paradigm of masculinity for his time, seems to have loved only males, Philip of France in particular (Boswell, 1980; Greenberg, 1988). As noted earlier, in feudal Japan, samurai were famous for their ultra-masculinity and for their erotic relationships with young male attendants (Leupp, 1995). The close association of *nanshoku* with the samurai probably contributed to the popular practice of male-male erotic relationships across all social classes in Tokugawa Japan. Across a variety of cultures, masculinity is usually associated with active, insertive sexual behavior, regardless of the sex of one's partners.

Evidence that ultra-femininity was associated with women who love women in any culture is harder to find, largely because the conventional female gender role demanded that women be invisible and passive. Women who did not bow to male authority or who were sexually assertive were not considered *feminine*. Furthermore, historical accounts of women's sexuality are rare, except from a male's perspective. Still, there are examples of women-loving women, consistent with conventional female roles, in which same-sex eroticism is not viewed as unfeminine or masculine. Among the Lesotho of southern Africa, an older married woman and young unmarried girl engage in a long-term social and sexual partnership called a Mummy-Baby relationship (Gay, 1986). The Mummy-Baby relationship occurs independently of marriage and continues after the younger partner marries. In addition, in Mombasa, Africa, Swahili Muslim women, like many men in that culture, form age-structured relationships. The *basha* is an older married woman, and the *shoga* is a young single girl. However, unlike male *basha-shoga* couples, female couples continue their relationship after the younger partner marries, sometimes for life. Aboriginal Australians also form age-structured same-sex relationships, independent of marriage (Roheim, 1933). Anthropologist Evelyn Blackwood (1993) suspects that same-sex erotic relationships between women were common in sex-segregated cultures and polygynous households, although there is little historical record to support this hypothesis.

To summarize this lengthy section, contrary to constructionists' assertions, exclusive same-sex erotic attraction, at least for males, has been recognized across diverse cultures, although called by different names. What is more, certain sexual types, such as the mannish woman and effeminate man, have persisted throughout history. Constructionists have also failed to recognize and explain similarities in sexual types across cultures. In short, sufficient evidence suggests that many pre-industrial non-Western cultures were aware that people experience degrees of sexual attraction to each sex and that most men are capable of attraction to both sexes. Limited evidence suggests that many women are capable of sexual attraction to both sexes, although far fewer experience an exclusion attraction to one sex. Rather than stating that each culture constructs sexual types of people from scratch, it is more accurate to say that in many cultures sexual labels solidify, define, and limit commonsense ideas about people.

3.1.2 Knowledge and Experience Is Language-Based

A second core constructionist principle extends the idea of relative and subjective knowledge. That is, what we know and what we believe to be true is based in language (Berger & Luckmann, 1966). Social concepts and sexual categories, as well as knowledge about physical objects and events and their relationship to the world, vary by culture because words do not represent objective reality but refer instead to other socially constructed signs and symbols (Burr, 1995). In fact, language *constructs* social reality by providing us the means to organize ideas and communicate with each other (Hacking, 1986). Persistent linguistic signs and symbols within the constant discourse of the social world produce persistent behavioral roles, such as *husband* and *wife* or *police officer* and *judge*; enduring institutions, such as *marriage* and *government*; and our experience and knowledge of reality (Berger &

Luckmann, 1966). These persistent signs and symbols become *real fictions* in that they are constructed socially but treated as essential facts. Consequently, we know what is permissible in our social experience, because that which is possible is relative to how the world is ordered linguistically (Sapir, 1947). For example, conventional marriage discourse structures the world into those people who are married and those who are not. Marriage is a privileged social position that permits special knowledge—to know what it is like to be married—and social recognition—*we* rather than *I*—which encourages pity for the single person.   By contrast, legal discourse organizes social relationships as rights and privileges and defines people as enforcers or as subjects of the law. Legal discourse creates categories of people to reflect their relative position to each other, including their different civil privileges, and special knowledge of the law. *Judge, police officer, free citizen*, and *criminal* are such categories. And within a sexual discourse structured by social status, different erotic feelings and behaviors reflect one's role as *master* or *slave, husband* or *wife, lover* or *beloved, citizen* or *alien*, and *adult* or *youth*. Each sexual role is one of a contrasting pair, and each carries with it an understanding of personal and social experience and what is real. Without the contrasting opposite, a single sexual role is meaningless.

Within a discourse of sexual identity, the contrasting sexual roles are "heterosexual" and "homosexual." Erotic feelings and sexual behavior are specific to each sexual role, which are intimately tied to gender roles. According to constructionist thought, erotic feelings or sexual behaviors that fall outside of a specified sexual role are ignored, suppressed, or redefined in order to be consistent with the role because "heterosexuals" do not, by definition, have same-sex erotic feelings and "homosexuals" do not have other-sex erotic feelings. By convention, "heterosexual" men are masculine and take the active sexual role, whereas "homosexual" men are effeminate and prefer the passive sexual role. On the other hand, "heterosexual" women are supposed to be feminine and passive, whereas "lesbians" are butch and sexually aggressive. Because sexual roles are tied to gender role behavior, violating a sexual role means simultaneously violating what it means to be a *man* or a *woman*.

Like *husband/wife* and *lover/beloved*, the sexual pairs "heterosexual" and "homosexual" are interdependent; without the other, each is meaningless. The concepts are symbiotic; in referring to one, the other is implied. Socially, the "homosexual" is a placeholder for the "heterosexual" (Halperin, 1995). The "heterosexual" is the positive side of the coin; the "homosexual" flip side is the negative. As the negative term, the "homosexual" refers to all that the "heterosexual" is *not*—what people of a given sex should not be. In identifying as "heterosexual," people claim a privileged social position, while declaring at the same time what they are *not* "homosexual" (Foucault, 1978/1990; Katz, 1995; McIntosh, 1968/1990). Some research suggests that at least some men accept a "heterosexual" identity by actively *rejecting* a "gay" one (Eliason, 1995). Adopting a privileged sexual role by rejecting a socially stigmatized one is consistent with constructionist accounts of male heterosexuality.

Still, choice is a perception. In a world where the benefits for being "heterosexual" are so great and the support for heterosexuality is so pervasive, being "heterosexual" is a simple matter of conformity to social expectations. There are no real choices from a constructionist perspective. What is more, in a study of college students (Eliason, 1995), women, unlike men, reported that they first considered adopting a "lesbian" or "bisexual" identity before finally *choosing* a "heterosexual" one. Although for women sexual identity is more broadly defined than for men, women should never contemplate the fit of deviant sexual identify, if constructionists are correct.

For social constructionists who reject the idea of a biologic sexual drive (Foucault, 1978/1990; Plummer, 1981), humans are a *tabula rasa* of general pleasurable feelings and thoughts that beg the social world to direct and define what they feel and experience to be true. Human beings are products of the culture in which they live. Human culture is grounded in language—in social discourse. At this period in Western culture, sexuality is translated through abstract linguistic concepts of individuality, intrapsychic processes, personal happiness, public expression, and unalienable rights and freedoms. Thus, human sexuality is a linguistic idea and a real fiction (Padgug, 1979/1990). Only the dullness of language limits the kinds of sexualities and erotic relationships that are possible. Sexuality should be as unique and diverse as cultures are socially diverse. While sharing similar forms, sexualities should be dissimilar across cultures, according to constructionists.

3.1.2.1 Sexual Similarities

Critics of social constructionism may generally agree that *language defines experience and knowledge and shapes social reality* and that *sexual roles represent social positions relative to others.* However, critics do not find the unlimited diversity of sexual roles that constructionists predict. In fact, cultural variance in how sexual roles are structured is quite small. Dynes (1990) observes that of the approximately 5,000 societies on this planet the number of unique sexual cultures is tiny. There are perhaps a handful of unique ways in which cultures have structured sexual roles. Most writers have described these various sexual structures in the context of discussing same-sex male relationships, although similar or greater social constraints may influence the structure of male-female sexual relationships. Even so, for most cultures, sexual and erotic relationships, whether between the other-sex or between same-sex partners, are between *social unequals*. In most cultures, sexual relationships occur between people of different social statuses. In many cultures, sex is a marker of social status, and women rarely have rights or a status similar to men. Age is the most common marker of social status and a common means for structuring sexual and erotic relationships.

Social historian David Greenberg (1988) in his tome *The Construction of Homosexuality* describes four common forms of male relationships—transgenerational or age-structured, transgenderal or gender-structured, class-structured, and egalitarian. Herdt (1984/1993a) proposed similar categories of male-male erotic relationships. From my own reading of history and cultural anthropology, most male-female sexual relationships fit into the same categories. For the sake of disproving the constructionist claim that cultures produce an infinite variety of sexual relationships, it is worth briefly describing Greenberg's four common types of relationships.

Greenberg (1988) notes that *transgenerational* relationships are the most common form historically. In this case, partners are different ages: the age difference can span a few years or decades, with the dominant partner being older and male. In most cultures, adult men have greater social privilege and honor than youths of either sex and have greater economic resources, larger social networks, and political alliances. Adult(man)hood is a privileged position. One privilege often associated with male adulthood is the right to marry and have sexual relationships with social subordinates such as women, girls, or boys. The marriage of an adult man in his twenties or thirties to a young girl of age 11 or 13 is a transgenerational relationship. Samurai and their young male attendants in feudal Japan are examples of transgenerational same-sex male relationships (Leupp, 1995). Male paiderastic relationships in ancient Greece are other examples. The ideal *erastēs* was a free-born adult male in his twenties who took the active sexual role, and the ideal *erōmenos* was a prepubescent boy who took the passive sexual role (Dover, 1978/1989). By custom, the *erastēs-erōmenos* relationship ended when the beloved began to grow facial hair. *Erastēs-erōmenos* relationships first developed among the Achaeans and early Greek peoples as male initiation rites, similar to boy-inseminating rituals among bands in New Guinea (Dover, 1978/1989; Halperin, 1990). Because the practice of *erastēs-erōmenos* was based on affection, there were notable exceptions to the general rule of age disparity. As already noted, Achilles and Patroklos were both adult men and thought to be close in age. Pausanias and Agathon were also grown men. In fact, Boswell (1980; 1990) argues that the age disparity between males in *erastēs-erōmenos* relationships in ancient Greece was often small. Not much is known about the age of partners in same-sex female relationships. However, Sappho of Lesbos, who lived between the seventh and sixth centuries B.C.E., was known to have intense erotic desire for schoolgirls at the *thiasos* where she taught (Dover, 1978/1989). Sappho's use of language in her poetry was similar to that of male *erastai* to their *erōmenoi*. It is likely that Sappho's female relationships reflected the age-structure of other sexual/romantic relationships at the time.

Greenberg's second common form of sexual and erotic relationship is *transgenderal*. Here, partners are different genders or play different gender roles, although they are not necessarily different sexes. Male-female sexual relationships are between different sexes and different genders. However, partners may also be between a conventionally sexed individual and a cross-gendered or *third*-gendered person, meaning someone whose identified gender is different than their morphological sex. Within a gendered conceptualization of sexual relationships, *men* penetrate, and *women* are penetrated. A male who desires to be penetrated is not a *man*, by this definition, and a female who wants to penetrate is not a *woman*. Gender is not considered fixed and invariant; therefore, gender roles are important in demonstrating who one is. Latin and Mediterranean cultures demonstrate this type of sexual structure, where passive or effeminate males are *not* men (Carrier, 1980; Lancaster, 1995). *Men* are loud, controlling, tough, and hypersexual, whereas *women* are passive, compliant, and focused on the family. Masculinity does not prohibit same-sex erotic activity, as long as *men* take the active insertive role and do not engage exclusively in such behavior. Males who are penetrated orally or anally are *not* men.

Among traditional Native American Indian cultures, males or females who exhibit cross-gender characteristics may comprise a *third gender*. Many Native American Indian cultures recognize third-gendered individuals (Bleys, 1995; Greenberg, 1988; W. Williams, 1986). Early French explorers first referred to Indian males who dressed like women and enjoyed the intimate company of men as *bardash* and later, *berdache*, which loosely translates as "sodomite" (D'Emilio & Freedman, 1988/1997; Greenberg, 1988). These individuals were thought to represent a spiritual bridge between men and women and are now known as *two-spirited people*. Female *two-spirited people* also cross-dressed, performed male activities, and even married women. (For more information on this subject, see discussion in chapter 2). Other cultures that recognized third genders include the Meru of Kenya, the Iban (Sea Dyak), late 19th century Sarawak in northwestern Borneo, the Ngadju Dyak of southern Borneo, Bugis of South Celebes, and the Pelow Islands (Greenberg, 1988). The *mahu* of Tahiti wore female clothing and engaged in sexual relations with men, although, at least in 18th century Tahiti, same-sex erotic contact was generally common among males. Among the Kwayama of the Angolan Bantu, cross-dressing third-gendered males served as secondary wives to men (Greenberg, 1988). The Nandi of Kenya; Dinka and Nuer of Sudan; Konso and Amhara of Ethiopa; Ottoro of Nubia; Fanti of

Ghana; Dongo and Ovimbundu of Angola; Thonga of Rhodesia; Tanala and Bara of Madagascar; Wolof of Senegal; and Lango, Iteso, Gisu, and Sebei of Uganda similarly utilized third-gendered males as secondary wives (Bleys, 1995; Greenberg, 1988).

A third common form of sexual and erotic relationship identified by Greenberg is *class-structured*, which rests on differences between the partners' social class. Social class can be defined in a number of ways—by age, sex or gender, sexual roles, and civil status, although here Greenberg is emphasizing civil status. In most cultures, sex is an inherent marker of social class, with males being superior to females. Thus, most male-female erotic relationships in history have been between individuals of unequal social status. Civil status refers to one's social position in the community and the rights and privileges associated with that position. Examples of class-structured relationships are husband/wife, master/slave or servant, patron/prostitute, and man/boy or man/woman. Some cultures, such as ancient Rome, permitted male citizens to have (insertive) sexual relations with practically anyone who was not a male citizen, including adult male non-citizens, slaves, freed slaves, boys, girls, wives, concubines, and prostitutes. Men of low civil status such as freed men or youths could play the active sexual role only if their partners, whichever sex, were socially inferior. Women in most cultures have low civil status and little sexual freedom, although free-born females and noblewomen have found opportunity to initiate sexual relations with social inferiors.

The ancient Babylonian story of Gilgamesh and Enkidu, mentioned earlier, presents an example of a class-structured erotic relationship (Greenberg, 1988). Gilgamesh was the fifth ruler in the First Dynasty of Uruk (ca. 2700-2500 B.C.E.), and Enkidu was a barbarian. According to the legend, Enkidu is sent to kill Gilgamesh who had angered the nobility by sexually violating male youths of good birth. After a long fight, Gilgamesh and Enkidu develop a respect for each other and a close affectionate relationship. The pair are repeatedly described as husband and wife in the *Gilgamesh Epic*. Upon Enkidu's death, Gilgamesh weeps like "a wailing woman" and veils the body like a bride (Halperin, 1990; Pritchard, 1958). Although the poem describes no graphic sexual behavior between Gilgamesh and Enkidu, their relationship was clearly intimate, and numerous sexual puns suggest just how.

The Roman Republic provides several examples of class-structured sexual relationships. Roman males were raised to be conquerors in every aspect of their life, including sexually (Cantarella, 1992). This meant being the penetrator. Submitting to another individual was weak and emasculating; making another submit was virile. The Roman philosopher Seneca (ca. 55-40 B.C.E.) stated in his *Controversies* that "losing one's virtue (through sexual passivity—*impudicitia*) is a crime for the free-born, a necessity for the slave, a duty for the freedman" (Cantarella, 1992, p. 99). Although Romans greatly admired Greek culture, they rejected the institution of romantic *paiderastia*. Slaves, not noble youths, were the most frequent sexual objects for Roman men. Polybius, a Greek visiting Rome in the second century B.C.E., observed that most Roman men had male lovers who were slaves (Boswell, 1980). In the second century C.E., almost 400 years later, Tatian, a Syrian Christian, commented that Roman men "collect herds of boys like grazing horses" (McMullen, 1982). Roman men, like the Greeks, also enjoyed debating the superiority of loving women versus loving boys. In the *Amatorius*, moralist Plutarch, after lengthy dialogue, concluded that the love of women within marriage was best, although he did not condemn boylove (Cantarella, 1992).

As well as being age-structured, male *basha-shoga* relationships among Mombasans, which I described earlier, present further examples of class-structured erotic relationships. Similarly, female Mummy-Baby relationships among the Lesotho of southern Africa are both age- and class-structured. Boy-inseminating rituals in Melanesia (Herdt, 1984/1993a,b, 1991) might also be both age- and class-structured. However, the Kamula and Vanuatu (New Hebrides) man-boy pairs refer to each other as husband and wife, which implies more of a class disparity than age difference (Baal, 1934/1966; Davenport, 1977; Serpenti, 1965, 1984/1993).

Finally, *egalitarian* relationships are the rarest form of sexual relationship and more typical of contemporary Western culture, which prizes autonomy, individualism, and democracy. The hallmark of egalitarian relationships is that partners are close in age, socially similar, have relatively equal power, and share decision making; one partner is not subservient or owned by the other. Historically, romantic friendships between same-sex partners have been egalitarian. From the 17th century through the 19th century, romantic friendships were especially popular between middle- and upper-class women (Faderman, 1981/1998). The term *romantic friendship* was not coined until the 18th century and referred primarily to special affectionate relationships between women. The assumption at the time was that women in romantic friendships were emotionally intimate but not sexual with each other. However, the passion that these women experienced for each other often exceeded and overshadowed their feelings for their husbands and for other men. Although they were not "lesbians" in the contemporary sense, they appear to have had a primary erotic attraction and affection for other women. Social historian Lillian Faderman describes several women who maintained significant romantic friendships, including the famous Ladies of Llangollen. In 1778, when very young, Eleanor Butler and Sarah Ponsonby left their families after Sarah refused to marry a man, and they set up a home together. Butler and Ponsonby shared one bed, not uncommon at the time, and pooled their financial resources. The couple was respected and well known among literary figures of the day and celebrated by poets.

Much information about romantic friendships among women is drawn from diaries and love letters—letter writing was a popular means of communication and an art form in the eighteenth and nineteenth centuries. Although hyperbolic sentimentality was commonplace at the time, there is considerable circumstantial evidence to suggest that more than witty word play was involved in expressions of affection. In France, Madame de Staël wrote to her dear friend Madame Récamier, "I love you with a love surpassing that of friendship. I go down on my knees to embrace you with all my heart" (Faderman, 1981/1998, p. 79). In England, Elizabeth Carter and Catherine Talbot, both 18th century writers, made the unusual decision not to marry under the pretext of caring for sick parents. Although Carter and Talbot never lived together, they maintained an extraordinarily close relationship for many years. In their frequent letters, Talbot rebuked Carter when she remarked too intently on the beauty of other women (Faderman, 1981/1998). After Talbot's death, Carter developed a close affectionate bond with another woman, Elizabeth Montagu. In another case, 18th century poet Anna Seward, when a young girl, fell deeply in love with Honora Sneyd when Sneyd lived briefly with Seward's family. Over Seward's vocal disapproval, Sneyd left the Seward household to marry Robert Edgeworth. Although Sneyd was a frequent subject of Seward's poems, their relationship remained strained for years as a result of Sneyd's marriage. Yet, it persisted and, even 30 years after Sneyd's death, Seward continued to mourn and fantasize about how happy they would have been together.

In the late 19th century, romantic friendships among unmarried, financially independent women were sometimes called *Boston marriages* (Faderman, 1981/1998). Usually these women lived together and were very involved in feminism and culture. Henry James described a Boston marriage between Olive Chancellor and Verena Tarrant in his popular 1885 novel, *The Bostonians*, the characterization was probably modeled on his sister. Real-life American Boston marriages included that of Willa Cather and Edith Lewis, Annie Fields and Sarah Orne Jewett, and Emma Stebbins and Charlotte Cushman. However, so-called Boston marriages were also popular among feminists in Europe, and participants included Marie Corelli and Bertha Vyver, Alice French ("Octave Thanet") and Jane Crawford, and Rosa Bonheur and Nathalie Micas.

To a lesser degree, men enjoyed romantic friendships during the 18th and 19th centuries. Idealized male friendships flourished in the 19th century, especially as subjects of popular fiction (Martin, 1989). The male friendship movement in Germany in the late 19th century—ostensibly an attempt to legitimize male love—was an outgrowth of romantic male friendships (Oosterhuis, 1991). By the early 20th century, romantic relationships and marriage in the West were beginning to be conceptualized as free contracts based on love and affection between autonomous individuals. As a result, male-female erotic relationships became increasingly egalitarian. At the same time, sexual attraction was being viewed as a gender-role based intrapsychic sexual orientation—fixed and exclusive and central to personal identity— thanks to Freud's influence. Consistent with new Western values, erotic relationships structured by sexual identity have been egalitarian for the most part. All the same, "gay" men and "lesbians" have sometimes modeled their relationships and sexual roles on gender role complements, *butch/femme* or *top/bottom*, although this is hardly the norm. Feminists refute the claim that the roles of husband and wife are truly egalitarian in "heterosexual" relationships, although Western women enjoy more financial and social freedoms than do women in non-Western cultures (Faludi, 1991).

Except for the egalitarian-type, sexual/erotic relationships are based on role complementarity and maximizing reproductive capacity. Until relatively recently, sexual and affectionate relationships have taken surprisingly similar forms. Advances in Western reproductive medicine and contraceptives have separated sexuality and sexual relationships from reproduction. Yet for the bulk of human history, why do we not find a few thousand or even several hundred very different forms of sexual relationships, as social constructionists hypothesize? Constructionists complain that biomedical theorists fail to account for diverse sexual cultures. However, constructionists fail to account for or even recognize the limited dissimilarities and remarkable commonalities across sexual cultures. Sexual relationships have not taken an infinite number of forms, but rather very specific forms. One major reason for this theoretical blindspot in social constructionist thinking is their preoccupation with the political function of sexual categorization, resulting in a perception of sexuality that is largely non-sexual (Wieringa, 1988) and nonbiological.

To summarize this section, constructionists have argued that language and culture create unique sexualities, making them dissimilar and removing development of sexual attraction from any uniform biological influence. They argue that diversity among sexual cultures supports this contention. However, other writers have found remarkable consistency across cultures and little diversity in the form that sexual relationships take. Until modern times, sexual relationships have reflected complementary gender roles and concern about reproduction. Thus, although language may indeed structure experience and knowledge, other factors besides language alone appear to structure or limit sexual attraction and behavior.

The relationship between language and sexual attraction comes up again later in this text. Several interactionist theorists of sexual attraction, including Kauth and Kalichman (1995), Money (1988), and Weinrich (1987a), describe the development of sexual orientation as analogous to language development. Toddlers have a unique ability to acquire and produce language at a remarkable rate and speed which suggests that this capacity at a particular age, given particular social conditions, is predisposed or "hardwired" in the brain. Similarly, interactionists suppose that humans are "hardwired" for sexual attraction but that the direction and meaning of erotic feelings are determined by social conditions. These ideas are explored in more detail in chapter 5.

3.1.3 The Observer Is Never Neutral; Observations Are Not Objective

Social constructionists insist that knowledge, as a product of language, is grounded in the values and historical period of the culture (Burr, 1995; Gergen, 1985). Implicit cultural values are not only reflected in knowledge but shape the pursuit of knowledge. Cultural values influence what questions researchers, scientists, moralists, and philosophers ask and how they conceptualize their findings. French philosopher Michel Foucault (1978/1990) described the effect of cultural values on intellectual thought as historical *epistemes* or epochs of characteristic thought and beliefs that are separated by transitional periods. Linguistics and philosophy professor Thomas Kuhn (1962/1996) noted that Science is also characterized by episodes of *paradigmatic* thought that are overturned eventually by accumulative anomalies. The shifting from one paradigm to another he called *scientific revolution*. Both Foucault's *epistemes* and Kuhn's *paradigms* refer to collective cultural assumptions about they way things work, to which most of us—including scientists and philosophers—conform until the underlying assumptions crumble under pressure to explain differences in experience. Thus, in their pursuit of knowledge observers bring with them a set of implicit cultural or paradigmatic biases about what is important and what they should find. Social constructionists argue that implicit biases prevent neutrality on the part of the observer. What is more, they assert that the observer often ends up confirming their implicit biases. Therefore, according to constructionist thinking, Science is hopelessly flawed, and scientists are as guilty as politicians in pursuing a biased agenda.

Cultural observations and anthropology texts are full of observer's assumptions about "primitives," nonwhites, women, and sexuality. Again, Evelyn Blackwood (1993) noted that in keeping with Western ideas about sexuality early anthropologists often pathologized same-sex erotic activity among non-Western cultures. Herdt's (1981, 1984/1993a) initial mislabeling of boy-inseminating rituals in New Guinea as "ritualized homosexuality" and "homosexual" behavior is a good example of how Western cultural assumptions infect even careful, considerate, thoughtful observations of non-

Western cultures. In a dramatic historical example of biased interpretation of culture, Spanish Conquistadors labeled the indigenous people of the Americas "sodomites" and executed most of them (Bleys, 1995). Catholic missionaries and explorers believed that same-sex erotic activity was sinful and warranted cruel death. Of course, the Spanish invaders also believed that they were morally superior to the primitive native Americans, and their open sexuality provided a handy justification for the main motive of the Spanish—to loot all material resources from the New World and obtain slaves.

Recorded history is a product of a particular point of view that engenders assumptions about reality—assumptions that change over time, with changes in the interpretation of history. Most historical writings, for example, represent the point of view of the educated writer, the ruling class, the winning side, or the dominant political position (Dynes, 1990). None of these people are objective. Alternative versions of reality by minority social and political factions or by those who were defeated by force are rarely recorded. A few decades ago, standard high school American history texts rarely mentioned women or people of color, except in supportive roles to white males or demeaning ways. Today, as a result of pressure from feminists and ethnic minority groups, American history texts mention the accomplishments of prominent women and African Americans, although from a majority culture point of view.

If knowledge reflects cultural values, then *who* produces that knowledge is important. The observer is not only shaped by native cultural beliefs, but the observer also reinforces and shapes those beliefs by influencing the social discourse (Foucault, 1978/1990). According to social constructionists, information about the world represents the observer's point of view, and particular points of view—particular discourses—favor certain individuals or social groups. The power and position of a speaker or a group is enhanced or maintained by convincing other members of the culture that a particular interpretation is valid and true (Burr, 1995). In short, to have the power to construct reality is to be heard and believed. One of the greatest achievements of social constructionists has been to refute scientific objectivity. Social constructionists have effectively demonstrated that scientists have used science to confirm assumptions about the social world and to reinforce their own authority. According to constructionists, by accepting biomedical explanations that people are born with certain sexual traits, we accept a particular version of "reality" and reinforce their ability to define people and the world. Thus, say social constructionists, "Truth" is largely political propaganda employed by groups to achieve or maintain social power. Controlling what is true is power. Historians who omitted information about an individual's same-sex erotic orientation and relationships or even fabricated "heterosexual" events participated in promoting a heterosexist version of reality (Norton, 1997). The result is that heterosexuality appears to be universal and normative, and all other affectionate relationships seem unusual and bizarre by contrast. For this very reason, heated debate erupts over *which* individuals and *what* version of historical events are depicted in history texts. Control over "reality" is at stake. *Revisionist* historians are interested in more than sexuality.[2] In attempts to control reality, various revisionists have asserted that the Holocaust never happened, Columbus was not looking for new trade routes but wanted to bring Christianity to the Indians, and the American Civil War had nothing to do with slavery (Herman, 1997).

Although social constructionists have been convincing in their assertions that scientists have a political agenda and seek to preserve their particular version of reality, they never acknowledge their own participation in the propaganda of Truth-knowledge and the pursuit of power. No constructionist seems to have openly acknowledged that his or her *goal is to persuade readers that the constructionist view is the correct one!* Indeed, most present their conclusions as rational (even objective!) criticism of dangerous social practices after thorough review and thoughtful analysis of the "facts." After constructionists blast biomedical theorists, the only view left standing is their own. Social constructionism is adept at deflating conventional positions and promoting clever objective-appearing discourse. However, for social constructionists to claim in the language of altruist and rational discourse that biomedical theorists manipulate language and social "reality" in order to convince others of their authority—but that constructionists are somehow *not* doing the same thing—is quite a rhetorical trick. As social constructionists are quick to point out, the ability to *persuade* is power over social reality, and truth is what has popular appeal (Berger & Luckmann, 1966; Gergen, 1985). Truth is not what is objectively valid.

3.1.4 The Power To Name Things

According to social constructionist theory, a coveted feature of the authoritative speaker is the ability to name things and, thereby, not just shape a phenomenon but also bring it into existence (Weeks, 1977, 1981, 1985). A label carries with it certain linguistic baggage that sets particular contextual and ideological boundaries for a phenomenon, creating not just a definition but more importantly a social image (Burr, 1995). Once a label is accepted, alternative definitions and ideologies are excluded (Hacking, 1986). Social constructionists argue that medical professionals in the late 19th century invented "homosexuals" in just this way (Weeks, 1977). Much social constructionist writing about sexuality is devoted to the creation of "homosexuals" during the past 100 years.

Naming a phenomenon does not occur in a vacuum; it takes place within the context of converging social forces (Hacking, 1986). "Homosexuals," for example, were not randomly or magically created out of air with the wave of a scientist's wand. They sprang out of a rich convergence of beliefs, values, and emerging social groups. The Industrial Revolution set the conditions for the birth of "homosexuals." This age brought tremendous economic prosperity to the West and prompted vast urban expansion, widespread acceptance of capitalist ideals, growth of the new middle class, redefinition of the family, and fears among the middle class about future productivity because of inherited degeneracy and contamination of the Caucasian race (Greenberg, 1988; Katz, 1995). The Western industrial economic boom also brought waves of immigrants, as well as disease and strange customs, stimulating much fear from naturalized Americans. The rapidly growing middle class fought to enhance its own respectability and distinguish itself from the filth and disease of the lower classes and from the decadence of the upper class (D'Emilio & Freedman, 1988/1997). Middle-class citizens set themselves apart morally and reinforced their hope of producing healthy children for the next generation by imposing strict behavioral limits on thought, word, and deed. The *social purity movement* of the 19th century was an outgrowth of these efforts. The popular social purity movement attempted to regulate what was perceived as the source of social degeneration—drinking, prostitution, venereal disease, and masturbation. At this time, same-sex erotic behavior was not a practice unto itself but was considered to be a form of excessive masturbation (Bullough, 1976). Such non-reproductive sexual activities as masturbation and same-sex erotic activity were thought to waste precious resources, lower work productivity, and promote genetic inferiority among the white race.

Michel Foucault (1978/1990) asserts that during the late 19th century Western culture moved from an emphasis on *external* social control to self-control. Guilt, fear, shame, and desire for acceptance as *normal* became the new means for society to regulate public and private behavior; even private behavior—like masturbation—was thought to have public consequences such as decreased fertility or lowered work productivity. In the late 19th century, excessive masturbation was thought to contribute to moral degeneracy, nervous shock, same-sex erotic behavior, and even death (Graham, 1848; Kellogg, 1882/1974). Within this swirl of social forces, medical practitioners, many of whom were the new middle class, attempted to establish their social position as medical and moral authorities by identifying intrapsychic disease and naming what is *normal* (Foucault, 1978/1990). The identification of mental or psychological sickness and *normal* behavior was a new and significant foray for physicians. In the late 19th century, physicians declared that sexual perverts—excessive masturbators and that new group of people called "homosexuals"—were unlike and distinguishable from so-called *normal* people. According to these new experts, such people could be identified by a host of unpleasant symptoms, including cloudy thinking, coughing, fevers, headaches, pimples, gastrointestinal distress, violent cramps, and constipation (Bullough, 1976). Despite the obvious fact that these diagnostic indicators were vague and unremarkable, *normal* people were assured that they were not a "homosexual" pervert—although the occasional headache, upset stomach, and cough probably maintained a lingering worry about latent homosexuality.

In one of the first social constructionist essays on the origin of sexual attraction, sociologist Mary McIntosh (1968/1990) hypothesized that the invention of the "homosexual" in the late 19th century functioned to alleviate middle-class fear of social contamination. By separating out deviants from *normal* people, (the valuable members of) society remained pure and protected. McIntosh also claimed that once the label "homosexual" was available to characterize individuals who engaged in same-sex erotic activity, sexual deviants began to internalize the label and thus became, for all practical purposes, "homosexuals." In other words, once the label was applied, these individuals fulfilled the prophecy by being "homosexual." According to McIntosh, sexual deviants became "homosexual" by channeling sexual feelings in the defined and expected direction and by taking on the supposed characteristics of "homosexuals," because that was how the experts believed they were *supposed* to behave. For society, the "homosexual" marked the boundary between *normal* and *abnormal* sexuality. Soon after the invention of the "homosexual," *normal sexuality*, thanks initially to Freud, acquired a label and role—heterosexuality and the "heterosexual" (Katz, 1995; Weeks, 1985). With the invention of these sexual categories, erotic feelings made up essential elements of personal identity; people were defined by the direction of their sexual feelings. Popular acceptance of these ideas and the identity labels solidified the authority of the medical and psychiatric professions and established a new species of individuals (Foucault, 1978/1990).

As noted earlier, to name a thing is to influence social reality, which is the kind of power that every social group desires, including physicians in the late 19th century. According to social constructionists, physicians exploited the ripe social conditions of the time to enhance their own position of authority through control over the identification of sexual perversion and *normal* sexual behavior. Similarly, other social groups have fought to define phenomena, and these groups greatly benefit when their version of the truth is accepted by popular culture. A few examples of this process will suffice. Until the 1970s, security agencies of the United States government—the Department of Defense (DoD), Federal Bureau of Investigation (FBI), Central

Intelligence Agency (CIA), and National Security Agency, for example—claimed that "homosexuals" were a significant threat to national security (Sarbin & Karols, 1988). Allegedly, "homosexuals" were emotionally unstable and more prone than "heterosexuals" to sexual blackmail. Consequently, known or suspected "homosexuals" were denied top security clearances, prohibited from holding sensitive positions, and denied employment by the federal government, although no credible evidence ever existed that "homosexuals" constituted a threat to the United States government or to any other government. Yet these government agencies received substantial funding and authority to defend the United States against threats and, without a threat, these agencies lose power or cease to exist—an unimaginable thought for supporters of these agencies. Thus, for many years, "homosexuals" were a believable, sufficiently scary threat to justify the need for tighter security. The DoD currently employs the specter of the "homosexual" to justify the loss of privacy for military personnel, regulate their sexual behavior, present an image of a masculine/anti-feminine military, and reinforce the perception of a "threat (to the military) from within." Again, a threat justifies a Department of Defense. In another example, during the 1930s to the 1960s, FBI Director, J. Edgar Hoover, who lived with his male partner and occasionally performed in drag in private, successfully exploited the threat of "homosexuals" for his own personal and professional gain (Signorile, 1993).

In 1977, Florida orange juice spokesperson, singer, and former Miss Oklahoma, Anita Bryant, used her celebrity and access to the media to raise the threat of the "homosexual" in order to redefine a nondiscrimination ordinance in Dade County, Florida, that included sexual orientation (D'Emilio & Freedman, 1988/1997). Bryant did this so effectively that she and a group of religious and political conservatives halted and, in several cases, reversed civil rights legislation in several locales across the United States. Bryant reframed the civil rights ordinance in Dade County as a pro-homosexual "assault" on helpless children. As a self-identified expert, brave Bryant confessed, "Some of the stories I could tell you of child recruitment and child abuse by homosexuals would turn your stomach" (D'Emilio & Freedman, 1988/1997, p. 347). She described "gay" men and "lesbians" as "human garbage" and formed Save Our Children, Inc. Contemporary political and religious conservatives employ a strategy similar to Bryant's by redefining antidiscrimination laws as "special rights" or as hiring quotas for "gay" men and "lesbians."

Attempts to control the definition of phenomena were common throughout the Whitewater and Lewinsky investigations and President Clinton's impeachment proceedings. Each evening on the national news, Republican and Democratic congressional leaders, representatives of the Clinton administration, and the independent counsel "spun" their versions of "reality" based on events of that day. Each version varied considerably from the others. Persuasively describing the events of the day was important to all sides because winning public opinion meant controlling social reality and gaining or maintaining political power; convincing the American people of a particular reality was the objective of the speakers, and the stakes were high.

3.1.4.1 Top-Down, Bottom-Up, or the Interaction

Critics of social constructionism accept that naming a phenomenon has political effects, but they opine that constructionists are completely wrong about the medical profession's creation of the "homosexual" (Dynes, 1990; Hacking, 1986; Norton, 1992, 1997). Social historian Norton (1992, 1997), for one, points out that a subculture of individuals with same-sex erotic interests existed in major European cities *at least 200 years before physicians used the term.* These individuals already had a terminology for themselves, which included *molly, madge cull, bugger, Ganymede, Gany-boy, queen, queer, punk, sodomite,* and *tribade,* to name a few (Norton, 1997). What is more, the term "homosexual" was only *adopted,* not invented, by the medical profession (Greenberg, 1988; Katz, 1995). Although medical and psychiatric professionals exploited the term, the clinical pathologization of homosexuality was not widely accepted by popular culture until well into the 20th century (D'Emilio & Freedman, 1988/1997; Norton, 1997). People with exclusive same-sex erotic interests identified themselves as such long before the term "homosexual" was recognized by society as a type of person. For much of the late 19th century, the term was confined to professional discourse in medical and psychiatric journals and texts, to which the public did not have access.

Social constructionists appear to take a "top-down" approach to identity labeling that views the labeled individual as a passive, helpless victim of powerful social forces (Dynes, 1990; Hacking, 1986). Yet considerable historical evidence suggests that people with same-sex erotic interests named themselves long before physicians named them. Ian Hacking (1986) argues that social forces and anomalous individuals *interact* to create new identities of people. Hacking suggests that new categories of people usually come into existence at about the same time as individuals begin to fit those categories in a process he calls *dynamic nominalism.* In other words, individuals who have set themselves apart from others and the social groups who look to exploit differences in others converge in the construction of new social categories of people. In giving singular credit to the medicopsychiatric profession, social constructionists fail to recognize the much earlier and persistent "bottom-up" contribution of those people who recognized their own same-sex erotic interests and had already labeled themselves. Thus, contrary to social constructionist thought, physicians in the late 19th century did not create the "homosexual;" they facilitated the popularization of the "homosexual" identity. While labeling homosexuality enhanced the social position of physicians, many other forces were also at work. Constructionists have overlooked the significant contribution of the people who identified themselves with the label.

3.1.5 Sexuality Is a Relational Position

Another important core principle of social constructionism is that sexuality, including the direction of sexual attraction, is a *relational* position (Foucault, 1978/1990; Padgug, 1979/1990). Said another way, how people interpret their erotic feelings and how they behave is relative to their social position compared to others. Several constructionists including Berger and Luckmann (1966), Foucault (1978/1990), Gagnon and Simon (1973), Plummer (1981), and Weeks (1985, 1988) view sexuality and the direction of sexual attraction as nonspecific and fluid. For them, the origin of erotic feelings is trivial and never explored, although both Plummer and Weeks imply the existence of a general biological sex drive. Because people are born not knowing how to be sexual or how to have a sexuality, they must be taught what feels *sexual*, what sexual behavior is appropriate, and when sexual behavior is permitted. Through their interactions with other people, individuals learn how to behave and know whether to interpret feelings as sexual; in other words, no one is sexual in social isolation (Burr, 1995). Without the presence of others, people have no sexuality, according to constructionists. Not even masturbation, a solitary sexual act, is independent of social context. Culture defines masturbation as a solitary and secretive sexual act that most people perform but should not discuss in public; masturbation is what one does sexually when alone—in the physical absence of a sexual partner. However, the sexual partner may be present cognitively in fantasy.

According to constructionist thinking, people learn at an early age that their role or position in a social situation defines what behavior is *sexual* and what is not. We all play multiple roles in social situations. In this culture, for example, children learn that one generally should not touch their genitals in public and that adults, outside of parents, should not touch children's genitals. Most adults are horrified when the boundaries of parent-child or adult-child relationship are violated by sexual behavior. Touching a child's genitalia within a parent-child relationship is permitted and considered non-sexual; however, the same behavior by a nonparent adult is sexual and criminal. One's social position in relation to others in the context defines the behavior. If Aunt Martha kisses her young nephew on the cheek, the behavior is seen as a customary affectionate greeting. A kiss on the cheek between adults may be viewed as sexual, depending upon their relationship. During a candlelight dinner, a kiss on the cheek between a man and a woman suggests sensuality and the potential for sexual intimacy. If the kissers are two unrelated men, the behavior may also suggest sexual intimacy. However, in Eastern European cultures, a kiss between two men may signify a warm greeting between old friends or new ones. A decision to kiss or to file a grievance of sexual harassment depends upon the relationship between the participants and their social context. As the roles between people change, behavior and the interpretation of feelings also change.

What social constructionists say is that *who* people love depends on *what* social roles they participate in; *sexual identity* is a social role that influences whom people love (Blumstein & Schwartz, 1990). People who accept a "heterosexual" role love members of the other sex and find them sexually attractive. Their social role demands this. On the other hand, loving someone of the other-sex may facilitate acceptance of a "heterosexual" identity. Philip Blumstein and Pepper Schwartz (1990) argue that sex drive is not sex-specific and that sexual attraction is fluid. They claim that who people love reinforces or challenges sexual identity labels, which are fictional social conventions anyway. In other words, conventional sexual identities limit which sex people love; the few people who allow social roles other than sexual identity to guide their sexual feelings fall in love with the individuals and thus appear to change their sexual orientation. According to Blumstein and Schwartz, women are more likely than men to fall in love with a person rather than a gender and, consequently, appear to have a fluid sexuality. Blumstein and Schwartz predict that men and women would have more varied kinds of romantic relationships if society did not saddle them with sexual identities.

Although social roles within a given context certainly influence meaning and sexual behavior, critics claim that social constructionists go too far. For instance, constructionists largely conceptualize people as passive objects of sexual discourse who exert little influence on their own sexual identity (Dynes, 1990), *except* when they decide to change their sexual identity (Blumstein & Schwartz, 1990). They also believe that sexual attraction is completely nonspecific to gender despite historical, contemporary, and cross-cultural evidence of exclusive sexual attraction among some individuals but not others (Norton, 1997). Social constructionists have not well described how or why individuals participate in the internalization of negative sexual identity labels, as McIntosh (1968/1990) suggests. It is wholly unclear from constructionist writings why some people who engage in same-sex erotic behavior identify as "homosexual," while others who engage in the same behavior avoid internalization of the "homosexual" label and identify instead as "heterosexual." The hypothetical social conditions that permit such variability are never discussed. Personal traits are not the answer because constructionists never account for them. It is equally unclear why a few people resist or change long established sexual identity labels and move from a socially accepted "heterosexual" identity to a disapproved and problematic "homosexual" one. Blumstein and Schwartz's (1990) claim rings hollow. How could the social benefits of a stigmatized identity ever be so powerful that one would give up a socially privileged identity? Furthermore, if the direction of sexual attraction is truly open to definition, it is very difficult to understand why "homosexuals" do not all become "heterosexuals." The social advantages are far greater. I find it hard to believe that "homosexuals" remain "homosexuals" simply because they have fallen in love with someone of the same-sex. If that were the case, it seems reasonable that single "homosexuals" would find lots of incentives to love someone of the other-sex and become "heterosexual."

And finally, social constructionists never explain how individual determination and motivation exists within a socially constructed "reality" in which people passively conform to their social roles. Why do some people resist roles and reality rather than passively conform? *From where* does determination and the desire for personal advantage come? If these are socially constructed traits, constructionists have failed to explain how they develop. Such explanations are necessary to accept constructionist accounts of sexual attraction as a relational position.

3.1.6 Causal Explanations Are Narratives

One final social constructionist principle remains: Living in a world of subjective reality, the "cause" of social phenomena will never be within our grasp (Berger & Luckmann, 1966; Burr, 1995). Objective causes are not knowable; we have only our subjective experience on which to draw conclusions. Reasons provided for a phenomenon—even scientific reasons—are not a cause; they are narratives about a cause. Everyday people as well as scientists create narratives about causes. Personal narratives, for example, are stories we tell ourselves about how things work in this world relative to us. Rationales or causal narratives help people define the situation and control over reality by persuading others to believe what they believe.

According to social constructionists, scientific causes about the origin of sexual orientation and personal narratives about sexual identity development are no different. Both function to support the speaker's position and neither reveals the actual cause of sexual orientation. Constructionists note that personal narratives are constantly rewritten to support the individual's present circumstances and beliefs. As people incorporate new experiences and beliefs, what used to be "true" about oneself is shifted or modified or even fabricated. Social historian Jeffrey Weeks (1995) claims that personal narratives are meant to convince ourselves and perhaps others that our experience has led irrefutably to our present condition; personal narratives are not meant to document objective events.   Therefore, it is perfectly consistent that a "gay" person cannot point to a seminal childhood experience as producing or altering his or her sexual orientation (Berger & Luckmann, 1966).   Memory cannot be relied upon for objective data. Personal narratives are adapted and remembered in ways that are consistent with *current* social beliefs and politics. In short, they are a collection of diverse social experiences, smoothed over and integrated into a consistent story by memory and through the ongoing social discourse about oneself (Berger & Luckmann, 1966; Gergen, 1985; Weeks, 1995). When new social roles are adopted, such as *parent* or *homosexual* or *Nobel Prize winner*, the personal narrative is altered, and memories shift and change to support the new role. Significant contemporary events such as an assault or election to public office or gaining startling new personal knowledge (such as having been adopted) can send shock waves through our storehouse of memories and ignite a comprehensive reevaluation of past experiences and alteration of memories.

3.1.6.1 When is a Cause Not a Cause?

Because objective causes are illusive or even fictional, social constructionists largely avoid speculation about the origin of sexual orientation. In the social constructionist bible, *The History of Sexuality*, Foucault (1978/1990) refuses to speculate about the cause of homosexuality and instead describes a history of social discourse about homosexuality. Indeed, social constructionists, for the most part, are more interested in *how* social groups use information and ideas about sexual attraction to create social reality than in explaining the origin of sexual attraction.   The political function of sexual discourse is the primary interest of constructionists. Nevertheless, causal explanations about same-sex erotic attraction are assumed in constructionist writings without question. But, how valid are these implicit causes?

Mary McIntosh (1968/1990), considered by many gay theorists to be the mother of social constructionism, employs *labeling theory* to describe how individuals come to think of themselves as "gay" or "lesbian." Labeling theory enjoyed wide popularity during the 1960s (Plummer, 1981). McIntosh asserts that assignment of a "homosexual" label follows performance of a deviant sexual act. Presumably, the act of labeling occurs more than once, although McIntosh never makes this clear. Over a period of time, the "homosexual" label or role is internalized, the deviant individual accepts it and conforms to it, and he or she is now "homosexual" for all practical purposes. According to McIntosh, the individual's reaction to the "homosexual" label is critical to the process of identification. Some people see themselves as consistent with the label and accept it; others do not. Identification with the stigmatized label of "homosexual" might be facilitated by performance of a deviant sexual act that supports special categorization of the individual. McIntosh suggests that people who are "homosexual" have first committed a sexual act that warrants attention and a social label. Thus, in her mind, same-sex erotic behavior precedes "homosexual" identity.

Poststructuralist philosopher Foucault (1978/1990) suggests a similar theory of "homosexual" identity development. He notes that the 19th-century medical profession created convincing diagnostic labels such as the "homosexual" that served the needs of the medical profession to establish their authority and the needs of society to marginalize a group. Yet in actuality, clinicians diagnosed few "homosexual" men and "lesbians." According to Foucault, as Western culture emphasized internal control of social behavior through guilt and shame, members of society took on the role of diagnostician. Some individuals diagnosed or labeled themselves as "homosexual" and then conformed to the role. He is not specific about who was likely to accept a "homosexual" label or under what circumstances labeling occurred. Whether same-sex erotic behavior precedes or follows a self-diagnosis as "homosexual" is never explained. In addition, Foucault asserts that awareness of how social forces manipulate information and identities allows for social resistance. In other words, sexual identity labels can be resisted and challenged when individuals become aware that they have been social pawns. Still, Foucault never explains what personal characteristics make social resistance and personal redefinition possible. Although Foucault (1978/1990) and others (McIntosh, 1968/1990; Plummer, 1981; Weeks, 1985) often allude to motivation and determination as internal qualities that provide the means for identity change, the nature of these qualities and their origin are never explored.

As unintentional causal explanations of same-sex erotic orientation, both McIntosh and Foucault present weak theories. Contrary to McIntosh, many "gay" men and "lesbians" report being aware of same-sex erotic feelings *before* acting on them (Boxer, Cook, & Herdt, 1989, 1991).

Curiously, they state that awareness of same-sex erotic interests *precedes* sexual behavior, *and by several years*. Thus, labeling as "homosexual" does not seem to be the product of sexual behavior, as McIntosh claims. This account is more in line with Foucault's theory, although he has no explanation for the origin of same-sex erotic feelings. The second major criticism of McIntosh and Foucault's theories of homosexuality is that people are viewed as passive or willing participants in their sexual categorization, without explaining why. Foucault suggests that lack of awareness promotes conformity and through knowledge individuals are somehow able to resist labeling, although he also seems to imply the involvement of personality traits. Both theorists portray people as empty vessels, disconnected from their physical bodies, and disconnected from their evolutionary history.

As a causal narrative, McIntosh and Foucault's explanations lack specificity and empirical support. Constructionist theories about the origin of the "homosexual" are unconvincing.

3.2 LIMITS OF SOCIAL CONSTRUCTIONISM

To escape the stranglehold that social constructionism has placed on the field of sexuality, we must understand the strengths and weaknesses of the theory. First, the theory is a strategy for critical analyses; it is *not* a scientific theory. In fact, social constructionism makes a poor scientific theory. The goal of constructionists is to point out how information and concepts within social discourse support various social groups and particular versions of social reality. They cannot say what reality should be. Social constructionism is a relativist philosophy that holds that social narratives about reality have value to the people invested in them; beliefs have no objective value. For social constructionists, social beliefs that "gay" people are demon-possessed and should have holes drilled into their heads in order to release the evil spirits that reside therein is as valid an explanation as believing that there is a "gay" gene. The sociopolitical climate determines what beliefs are valued. When these beliefs are identified, constructionism can be used to facilitate social change, if change is desired.

Feminists, gay/lesbian theorists, social activists, and others who feel marginalized by society make up the largest group of adherents to social constructionism. For these followers, social constructionism represents a tool for social liberation. Ironically, social constructionism as described by Foucault (1978/1990) suggests that egalitarianism and democracy are futile because social power is never equally distributed. Proponents of social constructionism demonstrate a desire to uncover lies about conventional reality and be free of social manipulation. Ironically, again, such a desire suggests that some social constructionists harbor an ultimate belief in a hidden intrinsic character of being, despite the theory's assertion to the contrary.[3]

Second, social constructionists focus their vitriolic condemnation on single-factor biomedical theories of same-sex eroticism while completely ignoring contemporary interactionist theories. This approach is uncharitable and deceptive. Social constructionists have devoted hundreds of pages to criticism of antiquated single-factor biomedical theories of homosexuality. Contemporary sophisticated interactionist theories of sexual attraction are never mentioned, as if rejection of one biomedical model rejects them all. Paradoxically, social constructionists often follow their criticism of old biomedical theories with their own *single-factor social labeling theory of homosexuality*. Again, as a causal theory, the labeling model fails on several counts. Only general information is provided about how people are labeled and how they come to identify with a stigmatized label. While McIntosh (1968/1990) hypothesizes that a "homosexual" label follows same-sex erotic behavior, most "gay" and "lesbian" youths report recognizing same-sex erotic feelings three to five years on the average before they engage in same-sex erotic behavior (Boxer, Cook, & Herdt, 1989, 1991). What is more, self-identification as "gay" or "lesbian" may not occur for several years after initiation of same-sex erotic behavior, and many youths who engage in same-sex erotic activity never identify as "gay" (Remafedi, Resnick, Blum, & Harris, 1992). *Who* labeled these kids? *Why* do some youths identify as "gay," and others never do? Social constructionists cannot say and give no sound explanation for why "gay" men and "lesbians" willingly accept an imposed label that promises stigma, rejection, and discrimination. Identification with a stigmatized label is especially unfathomable if people possess an unrestricted range of sexual feelings. Just exactly what social constructionists mean by a "homosexual" role is unclear. At least today, there are as many or more differences between "gays" as there are similarities (Epstein, 1987). In fact, the lack of a common group experience or lifestyle is what has hindered the "gay" rights movement from evolving beyond a small, loosely organized, fractious and indecisive group of individuals.

A third problem with social constructionism is its omission of any discussion about bisexuality, while promoting the idea that the absence of sexual categories will result in diverse sexual relationships. Constructionists probably avoid discussion of bisexuality because it could easily be misconstrued as advocating an essentialist concept. Historical and contemporary examples make it plain that everyone does not have a bisexual capacity for eroticism; some people seem to have an exclusive and persistent erotic interest in one sex. Weeks (1985) has a dream that eliminating the sexual categories of "homosexual" and "heterosexual" will unleash a marvelous freedom for people to have various sexual relationships. Yet when speculating about the future, social constructionists continue to frame these unrestrained sexual relationships in terms of gender and sexual identity (Blumstein & Schwartz, 1990; Weeks, 1985). Why not organize sexual relationships by concordance or discordance of ear shape, hair color, skin color, degree of sexual pleasure, power relations, age congruence, or developmental stage of life? Apparently, for constructionists, uncategorized and unstructured sexual relationships are not the logical result of eliminating current sexual categories. Perhaps the weight of history has something to do with this. Until very recently in Western culture, sexual relationships were nearly always organized by complementary gender roles and age/class differences, consistent with effective reproductive strategies. In fact, several forces place constraints on the number of ways in which sexual relationships can be structured, including population growth, reproductive and contraceptive technology, sexual pleasure, free agency, commodification of sex, and moral and legal systems.

Finally, social constructionists take a disembodied, purely social view of human beings. In rejecting biomedical models of sexuality, they reject biology and portray people as "talking heads" who are disconnected from and uninfluenced by their physical body. Nor are people apparently influenced by the same biological and instinctual forces to survive and to reproduce that directs all other animals. Constructionists only discuss the *social* impact of sexuality (Dynes, 1990). According to social constructionists, people are unaffected or only marginally affected by biologic conditions such as normal genetic variation, race, sex, hormones, endorphins, neuromodulaters, environmental toxins, viral infection, disease, health, digestion, sensation, temperature, temperament, age, sensory perception, mood, or orgasm. Yet many of these factors contribute to individual variability in how social discourse is received, processed, filtered, linked to other information, interpreted, stored, retrieved, and utilized to initiate action. Although some social constructionists acknowledge a biologic "precondition" or "potentialities" of sexuality (Padgug, 1979/1990), they quickly add that biologic events are "never unmediated" (p. 51). This phrasing implies that social forces are more influential than biologic forces (Berger & Luckmann, 1966; Weeks, 1988) or that biology has only an initial impact on behavior and is not a continuous influence (Wieringa, 1988). Constructionists are not specific about what biologic preconditions are active on sexual attraction, because to do so might start the slippery slide into essentialism.

Furthermore, variety in sexual categories across cultures does not indicate that sexual orientation is uninfluenced by biology (Stein, 1990). A particular culture might ignore certain qualities and emphasize others. Several premodern cultures have described some men and women as having an exclusive or preferential erotic interest in the same sex (Boswell, 1982-1983; Brooten, 1996). In the past, people with exclusive erotic interests may have been rare or may have avoided attention because they were no serious threat to the established social order.

In brief, although social constructionists are very good at poking holes in conventional thinking, the theory offers little in the way of a useful conceptualization for the development of sexual attraction. The theory just does not match recorded experience and observations across cultures, and its predictions fall short.

3.3 SALVAGING SOCIAL CONSTRUCTIONISM

Despite pointed criticism of social constructionism, elements have merit and deserve attention. First, social constructionism may function best as an *investigational strategy* to enhance scientific research. By giving more attention to the politics of scientific discourse, as constructionists suggest, scientists can better understand their subjects and constructed reality.

Second, adoption of a less ideological middle ground constructionism allows some biological influence on sexual orientation (Vance, 1991). However, acknowledging that biology has some impact on sexual attraction and behavior is not the same as adopting an interactionist point of view. From an interactionist perspective, the biologic body represents the *interface* between biology and culture. It is within the body that stimuli are detected, information is processed neurochemically and stored, synthesized, retrieved, and utilized in the production of future behavior. In addition, the ability of the body to respond to social stimuli depends upon good health and a rich nourishing social environment. Processing social stimuli continuously affects the body, altering neural pathways and neurochemical responses. Meaning attributed to stimuli is strongly social, and reinforced responses strengthen neural pathways in the brain and increase the future likelihood of similar behavior. An interaction between biology, culture, and the individual's learning history stands in sharp contrast to the simplistic social constructionist and essentialist theories described in the chapter. Table 2 provides a comparison of these different approaches.

Before we begin to build an interactionist theory of sexual attraction, we must discuss one more class of theories regarding sexual attraction— biological theories. In the next chapter, conventional biological theories of sexual orientation are reviewed, and some new suggestive data that have implications for an interactionist theory of sexual attraction are discussed.

TABLE 2
Radical Social Constructionist and Biomedical Essentialist Positions Compared to
Interactionism

Social Constructionism	Biomedical Essentialism	Interactionism
"Facts" are social agreements. There is no objective Truth; there are only perspectives.	*There is an underlying order to the universe. Objective truth can be found.*	*General principles can be identified. There are more relative truths than objective truths.*
"Knowledge" is a product of the socially powerful. Scientists are self-serving.	*Knowledge is the product of empirical investigation.*	*"Knowledge" has many forms. Science is a tool to find reliable and verifiable data. Every group has an agenda.*
There are no causes, only narratives.	*There are proximate and ultimate causes.*	*There are proximate and ultimate causes.*
Observers are never neutral, and observations are not objective. Science is biased and hopelessly flawed.	*Trained observers can be objective. Science is objective and verifiable.*	*Assumptions underlie all observations and must be examined. Science is a self-correcting process. Constructionism is a rhetorical device.*
"Homosexuals" and "heterosexuals" are products of recent Western culture, although same-sex erotic behavior has always existed.	*Homosexuals and heterosexuals are natural kinds. They were unrecognized or named differently in the past.*	*"Homosexual" and "heterosexual" are unique ways of being. People with exclusive sexual attraction have always existed and were recognized in the past.*
Labels create people	*Categories identify natural kinds.*	*Labels may create or reflect kinds of people.*
"Sexuality" is culturally defined and unique to each culture.	*Sexuality can be objectively defined. Sexual cultures differ but have much in common.*	*"Sexuality" varies across cultures, although sexual cultures share many similarities.*
"Sexual attraction" is culture-specific and can be defined in an infinite number of ways.	*Sexual attraction is biologically based and comes in two forms.*	*"Sexual attraction" is a complex interaction between biology, culture, and learning history.*
People are "empty organisms." Identities and behavior are determined through social interactions.	*People have traits that shape or determine their identities and behavior.*	*Traits modify cognition and social behavior within a culture. Identities are shaped by all of these factors.*

CHAPTER 4: PARTS AND PIECES: EVOLUTIONARY AND BIOLOGICAL MODELS

All animals are equal but some animals are more equal than others. —
George Orwell, *Animal Farm*

The social brain not only senses, it also creates social worlds. — Leslie
Brothers, *Friday's Footprints*

Human behavior is neurochemically based because humans are
organic creatures. This is not to say that the human body is just a bag of
chemicals. It is an organized system of neurochemical reactions that permits
higher-order thinking, consciousness, volition, and altruism. Interconnected
networks of neurocircuits allow the body to respond to the environment and
act on it. Detection of stimuli, perception, cognition, memory, emotion states,
feelings, and behavior are functions of internal neural and neurochemical
processes. Internal sexual feelings and external sexual behavior are also
products of neural and neurochemical processes.

Examining the human body can tell us a great deal. Knowing the
chain of neurochemical events that produce the direction of sexual attraction
is significant; however, it is not particularly meaningful. The meaning of
sexual attraction is not understood when we understand biological processes.
Sexual attraction, no doubt, has a biological explanation, but the meaning of
those facts come from culture. The main reason that biological scientists have
investigated sexual orientation is because the culture accepts that same-sex
erotic attraction is problematic and needs an explanation (or a cure). As
social constructionists rightly point out, scientists represent their culture, and
their work reflects the values of that culture. Thus, the discovery of how parts
and pieces of neural systems produce different sexual attractions is interesting
but does not provide definitive answers. Debate about the meaning of varied

111

sexual attractions will take place in a larger sociopolitical forum, and scientists will be one group among many participants.

Because contemporary Western culture has problematized same-sex erotic attraction, scientists and others have speculated that the "problem" lies in biology. The culture has been concerned mainly with male-male erotic attraction, and most scientific efforts have focused on male homosexuality, ignoring other forms of sexual attraction. Conventional biological models have proposed that same-sex male eroticism—and, therefore, sexual attraction in general—is a function of particular genes and sex hormones. In this chapter, I review research findings from studies on genetics, hormones, neuroanatomy, cognitive functioning, and twin and family prevalence.

The biological models discussed here are more sophisticated than the simplistic version described by social constructionists in chapter 3. As well, the theories in the second half of this chapter are more comprehensive than theories in the first half. Unfortunately, the material presented here is highly technical, detailed, and demanding of the reader. Still, familiarity with conventional biological research and how these models compare to social models (discussed in chapters 2 and 3) is necessary to appreciate the intricacies and benefits of an interactionist model of sexual attraction, which will be developed in the next chapter.

As in the previous chapter, I begin by describing the assumptions that underlie conventional biological investigations of sexual orientation. Table 3 lists several basic assumptions and their implications for biological research. Biomedical writers almost never discuss such assumptions, although they are clearly evident in their models and research questions. Familiarity with these general assumptions will allow critical readers to evaluate the veracity of the research data. From my own extensive literature review, the most common troublesome assumption in biological research on sexual orientation conflates sex, gender role, and sexual attraction. As a result, discussions about sexual orientation, sex hormones, and sex become a mind-numbing tossed salad of gender terms. One follower of this unfortunate practice is pioneer sexologist John Money (1988) who refers to inter-linked sex-typed variations as gender transpositions. In Money's model, a lipstick "lesbian," for example, is cross-sexed—having the characteristics of the other sex. In this case, a feminine "lesbian" has a feminine sex but a masculine sexual attraction, an androgynous or ambiguous gender role and, perhaps, a masculine neuroanatomy. (The gender transposition model and its limitations are discussed again in chapter 5.) The problem with linking gender and gender roles—that is, masculinity and femininity—to sex hormones, anatomy, and sexual attraction is that complex phenomena are restricted to two dimensions.[1] As this chapter will illustrate, it is simplistic and misleading to reduce sexual orientation to two dimensions. However, given this assumption in conventional biological research on sexuality, it is not surprising that many studies produce ambiguous or conflicting findings or find no differences at all.

TABLE 3
Assumptions of Biological Research on Sexual Attraction

1. *Humans are animals; therefore, human sexuality and sexual behavior should be homologous to infrahuman sexuality and behavior.* Implication: *Rather than study human sexual behavior, which is difficult and subject to ethical constraints, observations from animal behavior can be generalized to human sexuality.*

2. *Sexual attraction and desire can be deduced from sexual (genital) behavior, but in the absence of genital activity, direction of attraction cannot be inferred.* Implication: *Sexual orientation is attributed to infrahumans and to humans based solely on sexual activity. Sexual feelings and fantasies in the absence of behavior are dismissed as valid markers of sexual attraction.*

3. *Animal sexual behavior is regulated by sex hormones. Since humans are animals, human sexuality is regulated by sex hormones.* Implication: *Human sexual behavior must be a function of sex hormones, and atypical sexual behavior must be a function of atypical sex hormones.*

4. *Heterosexuality is normal and, therefore, needs no explanation or investigation.* Implication: *"Heterosexual" attraction is not studied. On the other hand, "homosexuals" (and "bisexuals") are not typical and demand explanation. Atypical sexual attraction is sign that something has gone wrong.*

5. *Sex hormones make us men and women.* Implication: *Sex hormones both create and possess gender. Gender roles are attributed to hormones; the effects of androgens and estrogens are dichotomous and mutually exclusive. Sex differentiation is two-dimensional. Androgens produce male-typical (heterosexual) behavior; estrogens produce female-typical (heterosexual) behavior.*

6. *Gender is dichotomous.* Implication: *"Heterosexual" women should have no male-typical qualities, and "heterosexual" men have no female-typical qualities. Sex-typical behavior is synonymous with heterosexuality. Sex, gender identity, and gender roles are treated as synonyms. Same-sex oriented individuals are cross-sexed and, within same-sex relationships, one partner plays the male role and the other plays the female role. Men who like men are like women, and women who like women are like men.*

7. *Sex differentiation and sexual orientation occur early and simultaneously.* Implication: *Sexual orientation is a byproduct of sex differentiation.*

8. *"Heterosexuals" have normal genes and experience typical hormone exposure.* Implication: *"Homosexuals" have abnormal genes and an atypical hormone exposure.*

9. *Reproduction is the ultimate goal of sexual behavior, and reproductive sex is genital sex. Thus, genital sex is the most significant kind of sex.* Implication: *Investigate only penile-vaginal activities. Non-"heterosexual" non-reproductive sex is trivial or irrelevant. Phallic penetration is required for sex. Anal sex and vaginal penetration with a dildo are poor imitations of penile-vaginal intercourse.*

10. *Heterosexuality and homosexuality are mutually exclusive terms.* Implication: *"Heterosexuals" experience no same-sex eroticism, and "homosexuals" have no other-sex erotic interests. Bisexuality is a special case of homosexuality.*

Social constructionists point to such conceptual missteps in biomedical research when concluding that the pursuit for biological causes of sexual orientation is flawed and misguided; given, they say, that sexual orientation is not an essential quality, biomedical researchers should find only trivial or meaningless differences between "homosexuals" and "heterosexuals."   Worse, constructionists add, biomedical researchers, unaware of their conceptual biases, simply find what they expect to find; that is, a belief in fundamental differences between "homosexuals" and "heterosexuals" cues investigators to find a difference and inflate the significance of any difference. This argument has merit. Historically, enthusiastic biomedical researchers have exaggerated trivial and artifactual differences between "homosexuals" and "heterosexuals." There is no question that bad science has dirtied the study of homosexuality.

However, the misuse of a technique does not negate the value of the tool, appropriately applied, and confirming a prediction is not always a sign of bad science. Researchers often make discoveries that were not predicted. Many significant scientific accomplishments have been accidental. Conceptual uniformity and a perfect past record of avoiding harm to others are unrealistic, impossible standards for any dynamic discipline, including social constructionism.

Before we pick our way through numerous studies and details, one additional general remark needs emphasis to prevent this principle of biology from becoming lost: *biology favors diversity and variability*. In fact, variability rather than homogeneity or uniformity is the biological norm.[2]

As noted above, most biological research on sexual orientation has focused on male homosexuality. But, investigators have assumed that the cause of heterosexual sexual attraction is appropriate exposure to the sex hormones that make men act like men—that is, *androgens*. The thinking is that men who like men have been exposed to less androgen than "heterosexual" men. This is the *androgen underexposure theory*. By extension although to a lesser extent, women who like women are though to have been exposed to excess androgen relative to "heterosexual" women. This is the *androgen overexposure theory*. This particular school of thought about sexual orientation comes from data on non-human mammals, rodents and primates, as well as non-mammals such as birds. Human sexuality, for many reasons, is difficult to study. Because humans share many biological similarities with primates and other mammals, animal models of sexuality may perhaps reveal important information about human sexuality. Too often, however, biological researchers have minimized species differences and relied heavily on conclusions drawn from non-human mammals.

Even so, without comparative data on rodent sexual behavior and natural experiments with individuals who have genetic and hormonal variations, we would know little about human fetal development, sex hormones, or sexual behavior. Animal studies have helped scientists understand many aspects of human sexual development that cannot be studied directly.   Ethical   concerns   prevent   scientists   from   experimentally manipulating sex hormones or other developmental variables. Therefore, a society that is interested in understanding the biology of human sexual orientation must be content with statements of probability based on convergent circumstantial evidence from a variety of sources, some not directly related to typical human sexual behavior. For these reasons, being cautious and critical of biological research on sexual orientation is necessary. Sorting out what *is* known about the direction of sexual attraction from what *can be* known given ethical constraints and the limits of medical technology is difficult.

Following a review of the androgen under- and overexposure theories, I describe other areas of biological research that are less specific to androgen imbalances. These studies explore the effect of exposure to estrogen on sexual attraction, personality traits, immune response detection, formation of the inner ear, and sex differences as predicted by evolutionary theory. Genetic and family studies examine individuals who have a primary same-sex erotic orientation. Finally, I attempt to draw conclusions from this varied wealth of data.

4.1 ANDROGEN UNDER- AND OVEREXPOSURE THEORIES

In mammals, *genetic sex* is determined by the sex chromosomes. Mammal zygotes receive 23 chromosomes from each parent. Two of the 46 chromosomes are sex chromosomes. Typical males have XY sex chromosomes, and typical females have XX sex chromosomes.[3] Prior to sex differentiation, mammal embryos—regardless of genetic sex—have the potential to develop as either male or female. Mammal embryos have parallel internal reproductive systems and undifferentiated external genitalia. During development, genes on the Y chromosome program the male fetus to develop testes and manufacture a class of sex hormones called *androgens*. One particular androgen is *testosterone*. The presence of androgens has obvious effects and is said to "masculinize" the fetus because exposure promotes development of heavy musculature, the penis, scrotum, and testes (Goy & McEwen, 1980). Androgens also contribute to male-typical behaviors such as aggressiveness, high activity level, exploring, mounting, and penile thrusting in rats and other mammals. With testosterone production, the male fetus releases a hormone called *Müllerian inhibiting substance* (MIS), which "defeminizes" by triggering the atrophy of the parallel precursor of the female internal reproductive system. "Masculinization" and "defeminization" are thought to occur during slightly different critical periods. Without the timely presence of testosterone, the fetus is "feminized" and "demasculinized." Even with a Y sex chromosome, *if testosterone is absent* the fetus develops as female. Thus, genetic sex is not always congruent with *genital or phenotypic sex*—what the organism looks like. Fetal androgen exposure also affects brain development and produces distinct sex differences in form and function. (Data on brain development is discussed later in this chapter.) Unlike the duplicate internal reproductive systems, parallel sex-typed pathways do not exist in the mammalian brain (McKnight, 1997). Therefore, *exposure to sex hormones creates independent or overlapping "masculine" and "feminine" neural pathways in the brain.*

Androgen is not the only sex hormone that has "masculine" effects. *Estradiol,* an estrogen and metabolite of androgen converted from *dihydrotestosterone* through a process called *aromatization,* has "masculinizing" effects on rodent brain structures (MacLusky & Naftolin, 1981).[4] Extremely high doses of estrogen have "masculinizing" effects on behavior for female and non-androgen producing male rats that is similar to exposure to testosterone (Goy & McEwen, 1980). Fetal males and females may be protected from high levels of maternal estrogens by placental mechanisms. In brief, fetal androgen exposure stimulates male physical development, specialization of specific brain structures such as the *sexually dimorphic nucleus of the preoptic area* (SDN-POA) in rats, and male sex-typed behaviors such as fighting, mounting, and penile-thrusting. The conventional biological view of human male hormonalization is founded on this sequence of events, derived largely from rat studies (Money, 1988).

A *sexually dimorphic nucleus* (SDN) is a brain structure that exhibits distinct sex differences. In male rats, the SDN-POA is five to six times larger than in females (Breedlove, 1994). The SDN-POA is particularly sensitive to estrogenic metabolites from androgen during prenatal development and influences at least some aspects of sexual behavior. Males exposed to low doses of androgens during development have smaller SDN-POAs and exhibit fewer male-typical behaviors and more frequent same-sex erotic behavior such as mounting other males. Environmental conditions also influence male fetal androgen production. For example, maternal stress during pregnancy— restraining a pregnant rat under bright lights—produces large amounts of maternal stress hormones and inhibits androgen production in male pups. Male rats of stressed mothers have smaller nuclei of the SDN-POA and exhibit fewer male typical behaviors and a few female-typical behaviors (Anderson, Rhees, & Fleming, 1985). As a result, biological theorists hypothesized that male-male sexual attraction is a function of androgen deficiency (Dörner & Hinz, 1968; Dörner *et al.*, 1975; Money, 1970; Money & Ehrhardt, 1972). This is the basis for the androgen underexposure theory of human male same-sex erotic attraction.

To backtrack for a moment, among mammals, female development occurs in the *absence* of androgens (Money, 1988). Without androgen, the internal male reproductive system atrophies or is "demasculinized." Typically, genetic females (XX) manufacture *estrogen,* which "feminizes" the Müllerian reproductive structure and external genitalia. The female fetus develops a female-typical physiology, including ovaries, vagina, labia, and clitoris. However, the presence of estrogen is less obvious than is androgen to development and less well understood. At any rate, the absence of androgens and moderate levels of estrogen contribute to the development of female-typical behaviors such as nesting, nurturing, and sexually presenting behavior, called *lordosis*. When triggered by appropriate stimuli, sexually responsive female rats rotate their pelvis and arch the back to permit access to their vagina. The conventional biological model of human female hormonalization is derived from data regarding female rat development (Money, 1988).

A notable difference between humans and rats is that rats are prone to multiple births. This phenomenon allows researchers to study the effects of androgen exposure on female rat pups. Because rats usually carry litters, a female pup nestled *in utero* next to or between male siblings is exposed to testosterone, which partially "masculinizes" the female (vom Saal, 1989). This spillover testosterone produces male-looking anatomy in females, increased aggression, and frequent mounting of other females. What is more, when testosterone is implanted near the SDN-POA in newborn female rats, the treated rats as adults engage in sexual mounting more than three times as often as untreated females (Christensen & Gorski, 1978). These and similar observations led biological researchers to suppose that human same-sex female sexual attraction is a function of higher than typical exposure to androgen (Dörner & Hinz, 1968; Dörner *et al.*, 1975; Money, 1970; Money & Ehrhardt, 1972). The cause of androgen overexposure in human females is thought to be due to disease and the presence of synthetic hormones.

Before moving on to discussion of data on human subjects, I wish to note several problems with early studies of androgen exposure in rats, since these studies are at the core of conventional biological thinking on human sexual orientation. First, the SDN-POA in rats is not well understood, and research findings are sometimes contradictory. For example, although lesions in the preoptic area reduce male copulatory behavior, lesions in the SDN portion of the POA alone have little or no effect on sexual behavior (DeJonge *et al.*, 1989; Arendash & Gorski, 1983). The reason is unclear. Second, the androgen under- and overexposure theories are an overly simplistic explanation for a complex set of interacting biological and social variables. There is, for example, a great deal of variability in sex-typed characteristics among typically hormonalized rats, as well as for humans. The androgen theory suggests that individuals who experience same-sex erotic attraction are *cross-sexed*—pejoratively meaning that they have characteristics or experiences associated with the other sex—on a number of variables, including physiological development. When this theory is applied to people, "gay" men *should* be soft and effeminate, and "heterosexual" men *should* be rugged and "masculine." Likewise, "heterosexual" women *should* be delicate and feminine, and "lesbians" *should* be big and butch. However, there are no obvious morphological or behavioral differences between "homosexual" and "heterosexual" men and "homosexual" and "heterosexual" women. Third, the androgen under- and overexposure theories reflect a *disease model* of sexual attraction. The theory assumes that same-sex erotic attraction is something that has gone wrong, *without* evidence of pathology. An alternative view is that same-sex erotic attraction is a function of normal hormonal variation. And, lastly, the androgen theory dichotomizes sex differentiation of the brain and sexual orientation. Both processes could be viewed on a continuum in which sex differentiation in the brain and the direction of attraction are matters of degree rather than threshold events.

Despite these conceptual problems, the androgen under- and overexposure theories have generated a tremendous amount of data in five general areas, and researchers often refer to these findings. Familiarity with this research is important to grasp its limitations and seductive persistence. The five areas include congenital hormone variations, maternal stress, estrogen feedback effects, anatomic and neuroanatomic differences, and cognitive functional differences. *Congenital* refers to hormonal disorders that are present at birth but are not necessarily genetic or inherited. Under congenital variations, four hormonal anomalies are discussed: androgen insensitivity (AI), 21-hydroxylase deficiency or congenital adrenal hyperplasia (CAH) syndrome, 5-alpha reductase deficiency (5-aR), and 17-beta-hydroxysteroid dehydrogenase deficiency (17-bHSD).

4.1.1 Congenital Hormone Variations

4.1.1.1 Androgen Insensitivity

Androgen insensitivity (AI) is a recessive trait in which genetic males are partially or completely unresponsive to androgen because of faulty androgen receptors. AI individuals produce typical amounts of androgen but cannot respond to them; they may, however, respond to estrogens converted from androgens. Complete AI, which is very rare, results in a female-looking infant who is usually assigned and reared as female. Usually, AI individuals develop a female gender identity, do not menstruate at puberty, and are attracted to and marry men as adults (Collaer & Hines, 1995; Money, Schwarz, & Lewis, 1984). Observations of AI genetic males who were raised as females led Money (1969, 1988) to conclude that *gender identity*—viewing oneself as male or female—is flexible and malleable, at least until the second or third year of life.

German endocrinologist Gunter Dörner concluded that male homosexuality is a special case of AI, perhaps due to stress-induced maternal suppression of fetal androgen (Dörner *et al.*, 1991; Dörner *et al.*, 1975). Dörner hypothesized that reduced androgen levels in men led to "demasculinization" and a lower sex drive. In laboratory rats, males that received less androgen *in utero* "demasculinized," and male rats that were castrated at birth showed little interest in sex. However, to date, no studies have examined whether fetal androgen exposure differs for men who later identify as "heterosexual" or "gay," and several studies have found no differences in circulating testosterone levels between adult "homosexual" and "heterosexual" men (see Meyer-Bahlburg, 1984, for review). There is also considerable *within-sex* variability in testosterone levels that is not specific to sexual orientation. In fact, circulating testosterone varies to the point that statistical significance tests are unreliable measures of individual difference (Birke, 1982). Most circulating testosterone is bound to a protein called *sex hormone binding globulin* and is inactive (McConaghy, 1993). Only about 5% of circulating testosterone is unbound and biologically active. Said a different way, most men fall within a range of circulating testosterone that varies throughout the day, and men differ from each other on testosterone level. Adult male and female levels of testosterone vary by time of day, season of the year, age, relationship status, social performance, and frequency of sexual activity (Booth & Dabbs, 1993; Booth, Shelley, Mazur, Tharp, & Kittok, 1989; Brown, Monti, & Corriveau, 1978; Dabbs, 1990; Knussman, Christiansen, & Couwenberge, 1986; Tsitouras, Martin, & Harman, 1982; Udry, Billy, Morris, Groff, & Raj, 1985). Several researchers have noted that "gay" men are no more likely to experience hormonal dysfunction than are "heterosexual" men (Byne & Parsons, 1993; Meyer-Bahlburg, 1984).

In a thorough review of the literature on AI, Marcia Collaer and Melissa Hines (1995) found that most AI individuals develop a gender identity consistent with their sex of rearing and report a "heterosexual" erotic orientation. However, Gooren and Cohen-Kettenis (1991) reported one case of a genetic male with an incomplete form of AI, who received genital surgery to look more feminine. He later developed a male gender identity and erotic interest toward women, despite a much lower than typical level of androgens. Also consistent with this finding is evidence that males with low exposure to androgens and hypospadias develop a male gender identity and engage in male-typical behaviors (Sandberg *et al.*, 1995). *Hypospadia* is a common urogenital disorder in which the urethra terminates at various points on the underside of the penis rather than at the tip.

Apparently, ultralow levels of androgen in men do not result automatically in erotic desire for males, female gender identity, or feminine behavior, contrary to the androgen underexposure theory. Most AI individuals accept an assigned gender identity and express other-sex erotic attraction, which suggests that low androgen exposure neither determines same-sex erotic attraction nor prohibits other-sex attraction.

4.1.1.2 Congenital Adrenal Hyperplasia

Congenital adrenal hyperplasia (CAH) or *21-hydroxylase deficiency* is another recessive trait that results in sex hormone irregularities. In this case, the adrenal glands of genetic females manufacture unusually large quantities of androgen, resulting in partial "masculinization." Consequently, CAH women tend to be large and "masculine" looking. A few CAH females have been assigned and reared as boys (Money & Daléry, 1977). CAH females typically have small breasts, an enlarged clitoris, partially fused labia, and a shallow vagina. When identified at birth, CAH females are administered glucocorticoids to halt adrenal enlargement and stop excessive androgen production. External genitalia are often altered surgically to appear more feminine.

There are two forms of CAH: *salt-wasting* and *simple virilizing*. Both forms involve a cortisol deficiency and excessive production of androgen. The salt-wasting form of CAH signifies a serious electrolyte imbalance and markedly reduced levels of *aldosterone*, a corticosteroid. The simple virilizing form does not involve an electrolyte imbalance and includes low but adequate levels of aldosterone.

According to the androgen theory, women who experience too much androgen should look "masculine" and be attracted to other women. Although most CAH women report "heterosexual" erotic interest, several studies have reported an increased prevalence of "homosexual" and "bisexual" fantasy and behavior among CAH women (Dittmann, Kappes & Kappes, 1992; Money, Schwartz & Lewis, 1984). One study found that 44% of CAH women had desired or experienced same-sex relationships (Dittmann, Kappes, & Kappes, 1992). Dittmann and colleagues decided that insecurities related to genital appearance as the result of surgery were not important to sexual behavior. However, the study authors were not convinced that high levels of androgen alone caused same-sex erotic attraction in CAH women. They reasoned that, at most, extreme androgen exposure only *predisposed* CAH women to same-sex eroticism. For non-CAH women, the role of androgen in determining the direction of erotic attraction may be even weaker. A review of CAH studies by Collaer and Hines (1995) concluded that in general CAH women engage in more male-typical behaviors and less female-typical behaviors and more often experience same-sex erotic attraction than non-CAH women. More dramatic differences were found between salt-wasters than simple virilizers. However, Collaer and Hines point out that behavioral differences among CAH women could be due to chronic illness, side effects of cortisone treatment, reaction to repeated genital inspections, altered self-image, and other hormone irregularities associated with the disease.

Thus, early high androgen exposure does not produce same-sex erotic attraction in women. Consistent with this fact, "lesbians" are no more likely to have hormonal disorders than are "heterosexual" women (Byne & Parsons,

1993; Meyer-Bahlburg, 1984). Although circulating androgen plays a role in female sexual desire (Fisher, 1998; McConaghy, 1993), "lesbians" and "heterosexual" women do not differ in levels of circulating androgen (Downey, Ehrhardt, Schiffman, Dyrenfurth, & Becker, 1987).

In further contradiction of the androgen theory, *excessive* androgen exposure in males does not produce hypermasculine "heterosexuals." Cases of CAH in genetic males have provided inconsistent behavioral results for two reasons (Collaer & Hines, 1995).   First, CAH males do not seem to experience excessive prenatal exposure to androgens, which is typically high anyway. Second, excessive prenatal and neonatal levels of adrenal androgen may down-regulate testicular production of androgen through hypothalamic and pituitary mechanisms.

4.1.1.3 5-Alpha Reductase and 17-Beta-Hydroxysteroid Dehydrogenase Deficiencies

5-alpha reductase deficiency (5-aR) is the lack of the enzyme that converts testosterone to *dihydrotestosterone* (DHT), the particular androgen that "masculinizes" male external genitalia. 5-aR individuals have lower levels of DHT but typical to high levels of testosterone. As a result of the 5-aR deficiency, affected boys are born with feminine-appearing or ambiguous genitalia and are often raised as girls. The penis is usually small with labia-like folds of skin that contain immature testes. At puberty, rising levels of testicular testosterone complete male development and produce a fully formed but small penis and scrotum (Imperato-McGinley, Guerrero, Gautier, & Peterson, 1974; Imperato-McGinley, Peterson, Gautier, & Sturla, 1979). The 5-aR enzyme is not necessary at puberty for testosterone to have an effect. Still, 5-aR men are usually sterile, and some are impotent. This disorder was first noted in the Dominican Republic, as a product of inbreeding (Imperato-McGinley *et al.*, 1974; Imperato-McGinley et. al, 1979). The phenomenon is so common in the Dominican Republic that 5-aR boys are called *guevedoces* or "eggs (testes) at twelve." Although raised as girls for the early part of their life, reportedly, 5-aR boys are accepted as men after "masculinization" at puberty. What is more, 5-aR men typically report erotic interest in women, contrary to what the androgen theory predicts (Imperato-McGinley, *et al.*, 1974; Imperato-McGinley, Miller, Wilson, Peterson, Shackleton, & Gajdusek, 1991).

17-beta-hydroxysteroid dehydrogenase deficiency (17-bHSD) is the lack of the enzyme that converts *androstenedione*, the immediate precursor of testosterone, into testosterone. 17-bHSD individuals typically have lower levels of testosterone and DHT and elevated levels of androstenedione. Similar to 5-aR boys, 17-bHSD boys are born with feminine-looking or ambiguous genitalia and are usually raised as girls until puberty. After puberty, testicular testosterone completes "masculine" development. Likewise, 17-bHSD boys adopt male gender identities, male roles, and report sexual attraction to females. Collaer and Hines (1995) note that if androgens are important to development of sexual attraction at all, then androgens in general (rather than just testosterone) or low levels of testosterone may be sufficient for other-sex attraction.

Data regarding 5-aR and 17bHSD boys appear to contradict the androgen theory of sexual orientation. However, Money (1988) suggests that a lower level of testosterone may be required to "masculinize" the brain than is required to "masculinize" the body. Thus, the ultralow level of testosterone among 5-aR and 17-bHSD fetuses may have been sufficient to "masculinize" their brains and establish other-sex erotic attraction but not to promote "masculine" physical development. Most researchers also attribute the ease at which 5-aR and 17-bHSD "girls" become "men" to powerful social influences surrounding gender and attraction. Given that 5-aR and 17-bHSD infants do not look *exactly* female, it is unlikely that they are raised unambiguously as girls (Baker, 1980). Furthermore, given the frequency of these disorders in the Dominican Republic, it is likely that the community recognizes and accepts that some children who look mostly like girls—but not quite so—will later become men (Herdt, 1990). The existence of the term *guevedoces* suggests just such an expectation. In addition, the higher social status of men in the community is a further inducement for 5-aR and 17-bHSD youths to adopt a male identity. Yet outside of the Dominican Republic, most 5-aR individuals live as women, presumably have erotic interest in men, and marry (Johnson *et al.*, 1986). Is there less expectation in other cultures for 5-aR individuals to adopt a male identity and experience other-sex erotic attraction? Does this suggest that sexual attraction is not set prenatally? At most, the data on 5-aR and 17-bHSD individuals fail to support the conventional androgen theory of sexual attraction. At a minimum, only low levels of androgens appear to influence sexual orientation. The significance of these data has yet to be determined.

4.1.2 Maternal Stress

German endocrinologist Dörner and American sexologist Money advocated the theory that maternal stress contributed to same-sex erotic attraction in male offspring. As previously mentioned, Dörner demonstrated that under severe stress—being confined to small spaces and subjected to extreme sensory stimulation—pregnant rats produced higher levels of circulating adrenaline and corticosteroids (Dörner & Docke, 1964). Maternal corticosteroids may block fetal *adrenocorticotrophic hormones* (ACTH), which stimulate testicular testosterone production. Thus, blocking ACTH reduces levels of fetal testosterone, resulting in adult male rats that more frequently present themselves sexually to other males and show less sexual interest in females (Dahlof, Hard, & Larsson, 1977; Dörner, 1979). From this laboratory data, Dörner drew the unwarranted conclusion that human same-sex male erotic attraction is caused by maternal suppression of fetal testosterone and incomplete "masculinization." He further speculated that the prevalence of men with same-sex erotic interests should be more common after periods of extreme social stress—such as war—than during times of peace and prosperity. And indeed, Dörner reported that men born during the war years in Germany had a higher probability of being "gay" than men born prior to the war (Dörner *et al.*, 1980). Money (1988) adopted the maternal stress idea and extended it further by speculating that a pregnant mother's overindulgence in drugs and alcohol could also inhibit full "masculinization" of male fetuses and promote same-sex erotic tendencies.

Bailey, Willerman, and Parks (1991) tested the maternal stress theory of same-sex male attraction by asking mothers of "gay" and "heterosexual" men and women to describe and rate the frequency and intensity of stressors that they experienced during pregnancy. In sharp contradiction to the maternal stress theory, mothers of both "gay" and "heterosexual" men reported *similar* levels of stress across all stages of pregnancy. Mothers of "gay" men reported experiencing no greater stress than when pregnant with a "heterosexual" male child. However, mothers of "lesbians" unexpectedly reported experiencing significant great stress during their first and second trimesters compared to their pregnancy with a "heterosexual" female child. The relevance of this particular finding is unclear. The maternal stress theory had not hypothesized a relationship to same-sex female attraction, although this finding refutes the androgen theory of same-sex female eroticism. Bailey and colleagues, unfortunately, did not assess alcohol and drug use. However, no studies to date have connected excessive maternal drug use during pregnancy to same-sex male sexual attraction.

A brief thought experiment is sufficient to illustrate the absurdity of this line of reasoning. If, for instance, poverty is an extreme stressor, then at least the prevalence of "gay" men should be more common among the poor and disadvantaged ethnic groups. Yet available data do not support this supposition. In fact, most published studies of "gay" men cite an overrepresentation of mostly white, educated, affluent individuals. Although sampling low-income, less educated, and ethnic minority individuals is always difficult, writers and researchers have never suggested that "gay" men are more likely than "heterosexual" men to have poor parents. Money's hypothesized relationship between maternal drug use and same-sex erotic attraction in male offspring is a good example of how contemporary social values contaminate science. A common conceptual flaw among theorists in general is the assumption that *bad* outcomes are inevitably produced by *bad* events; conversely, *bad* events must be the cause of *bad* outcomes. Thus, the *a priori* belief that same-sex male sexual attraction is abnormal or unhealthy leads mistakenly to the prediction that unhealthy and harmful events such as substance abuse are implicated.

In short, data in this section point out that what is true for rats is not necessarily true for people. Animal sexual behavior is not equivalent to human sexuality. In particular, there is no reliable evidence that maternal stress affects same-sex male sexual attraction. The effect of maternal stress on female-female sexual attraction was not predicted by the theory, and its meaning is unclear. In all, the maternal stress data fail to support the androgen under- and overexposure theories of sexual attraction.

4.1.3 Estrogen Feedback Effect

Not to be dissuaded, Dörner and colleagues (Dörner *et al.*, 1975), from their earlier work with rats, reasoned that if "gay" men were partially "masculinized," they—like women—should react to estrogen surges, similar to what accompanies ovulation. That is, "gay" men *should* respond physiologically like women. When women experience high levels of estrogen, the anterior pituitary gland produces a spurt of *luteinizing hormone* that initiates ovulation. This response to estrogen is called the *estrogen feedback effect* (EFE) and does not occur in men. To test his theory, Dörner and associates gave doses of estrogen to 21 "gay" and "bisexual" men who were hospitalized for treatment of sexually transmitted diseases and other illnesses. Indeed, he found a small EFE among "gay" and "bisexual" men, and no EFE among "heterosexual" men. Dörner concluded that the EFE among "gay" men resulted from low exposure to prenatal androgens. Critics noted, however, that Dörner's sample size was small and that he overlooked the fact that half the "gay" men in his study did *not* show an EFE (Birke, 1982; Hoult, 1984). Only about one-third of the "gay" men demonstrated a positive EFE response which, compared with the EFE response in women, was very small. What is more, Dörner was unable to account for the larger *negative* EFE among "bisexuals." Even so, another group of researchers were able to demonstrate dramatic EFE differences in response to estrogen priming between 12 "heterosexual" women, 17 "heterosexual" men, and 14 "gay" men (Gladue, Green, & Hellman, 1984).

Endocrinologist Louis Gooren (1986, 1995) dismissed Dörner's and Gladue's conclusions altogether. He claimed that both failed to take into account typical fluctuation of hormone levels. Testosterone levels fluctuate within a certain range among men and throughout the day, being about 25% higher in the morning.[5] Age also must be taken into account because sex hormone levels are affected by age, with older men showing a decrease in testosterone and an increase in estrogen. Consequently, individual variability in typical testosterone levels would mask an EFE, if it even existed. Gooren

(1986) attempted to replicate and improve on Dörner's and Gladue's studies with a larger controlled sample. He found no significant EFE for "gay" men. To maximize the possibility of an EFE, Gooren injected men who had a hormonal response most similar to women with *human chorionic gonadotropin*, a strong endocrine stimulus. Still no EFE. In a more direct test, Gooren (1995) gave his sample of "gay" men *luteinizing hormone releasing-hormone* (LHRH)—a hormone that controls luteinizing hormone secretion in women—and again found no marked EFE. He concluded that small hormone changes among "gay" men in his sample (and among men in Dörner's and Gladue's studies) were due to falling levels of testosterone and variable testicular function, and not to an EFE. Hendricks, Graber, and Rodriguez-Sierra (1989) also failed to replicate Dörner's and Gladue's results.

Although Dörner has never been interested in studying "lesbians," his EFE theory predicts that women who love women—having been "masculinized"—*should* demonstrate less of a luteinizing hormone response to estrogen priming compared with "heterosexual" women. An extensive literature review reveals no studies that have investigated this hypothesis. Overall, EFE studies have not supported the androgen theory of sexual orientation.

Reminiscent of social constructionist warnings about bad science, Dörner's research provides excellent and frightening examples of how *a priori* assumptions predict conclusions. Dörner consistently finds what he expects to find—that male homosexuality is a neurohormonal pathology and that "gay" men significantly differ from "heterosexual" men. Other researchers with apparently different assumptions fail to find the same outcomes as Dörner or draw the same conclusions.

4.1.4 Anatomic and Neuroanatomic Differences

Studies of neuroanatomic differences are another potential source of evidence for an androgen effect on eroticism. Sexually dimorphic structures in the brain are thought to result from exposure to prenatal androgens. Size of certain structures and functional differences generally differentiate male and female brains. Women, for example, tend to have a larger splenium of the corpus callosum (de Lacoste-Utamsing & Holloway, 1982), larger massa intermedia, and larger anterior commissure than do men (Allen & Gorski, 1991). The *splenium* is the posterior portion of the corpus callosum, the tightly packed neural axons that connect the two cerebral hemispheres. The *anterior commissure* sits at the forefront of the tightly massed tissue of the corpus callosum. The *massa intermedia* is a bridge of neural tissue that lies between the left and right thalami. Neural asymmetry between the sexes has also been noted. Women are more likely to have a large *planum temporale* in the left hemisphere than in the right compared to men (Wada, Clarke, & Hamm, 1975). Biological researchers assume that sexual orientation, like sex differentiation, is a product of prenatal hormone exposure. The androgen underexposure theory predicts that men who are attracted to men have been exposed to less prenatal androgens than have "heterosexual" men. Consequently, "gay" men should have "feminized" brains similar to "heterosexual" women. In other words, "gay" men *should* show sex differences in brain that are more similar to "heterosexual" women than to "heterosexual" men. Furthermore, "lesbians" *should* have "masculinized" brains with sex differences that are more similar to "heterosexual" men than "heterosexual" women.

Researchers sought confirmation of these hypotheses by looking for a human brain structure that is analogous to the SDN-POA of the rat and then comparing the size of the structure among individuals with different sexual orientations. As discussed earlier, the SDN-POA of male rat controls some copulatory behaviors (Gorski, Gordon, Shryne, & Southam, 1978). The SDN-POA in rats, rich in androgen and estrogen receptors, is especially sensitive to androgen a few days prior to birth and following birth. A large SDN-POA in adult male rats is related to the frequency of female-mounting and penile-thrusting. A structure that is potentially similar to the rat SDN-POA is the *medial preoptic area of the human hypothalamus*. The medial preoptic area of the hypothalamus contains four groups of neurons, and the structure is thought to regulate some aspects of sexual behavior. Swaab and Fliers (1985) located a sexually dimorphic nucleus in the *third interstitial nuclei of the anterior hypothalamus* (INAH3) that was larger in men than in women. A large INAH3 was presumed to facilitate male-typical sexual behaviors, much like the SDN-POA does in male rats. In 1991, neuroscientist Simon LeVay examined this area of the hypothalamus in 16 "heterosexual" men and women and 19 "gay" men (but no "lesbians") to determine whether the INAH3 differed by sexual orientation as well as by sex. Unfortunately, the sexual histories of LeVay's subjects were incomplete, and sexual orientation was largely speculative. Still, LeVay reported that "gay" men and "heterosexual" women had an INAH3 that was similar in size and smaller on the average than the INAH3 of "heterosexual" men. No bigger than a pinpoint, the INAH3 in "heterosexual" men compared with women was two to three times larger. LeVay (1993) concluded that a large male-typical INAH3 reflected a sexual orientation toward women. Said another way, "gay" men had too few INAH3 brain cells to be sexually attracted to women, according to LeVay. Of course, the inverse relationship would also be true—that "heterosexual" men require a large number of INAH3 cells for sexual attraction to women.

Conceptually, LeVay's findings could present a problem. LeVay asserted that the INAH3 functions much like the SDN-POA does in rats; the SDN-POA regulates sexual behavior but is not known to control sexual desire. The ability to perform certain sexual behaviors is *not* equivalent to sexual desire. LeVay confuses sexual orientation and sexual behavior. As motivated human behavior, erotic feelings are more complex than reflexively thrusting a penis into an opening or presenting one's backside to be penetrated. Yet, by comparing the human INAH3 to the rat SDN-POA, LeVay emphasizes the importance of sexual behavior over sexual feelings and implies that mechanical behavior is the *sine qua non* of human sexuality. In actuality, the mechanics of sexual behavior—mounting, thrusting, rubbing, fondling, kissing, licking, and sucking—differ little across individuals of various erotic interests. Only the sex of the desired partner varies. Therefore, it is unlikely that the human INAH3 is functionally similar to the rat SDN-POA. LeVay's premise is faulty, and his conclusions are suspect. His findings could easily reflect the different environmental experiences of his subjects rather than reveal anything about sexual orientation. No meaningful conclusions can be drawn from LeVay's study.

Besides INAH3, other sexually dimorphic nuclei have been implicated in the direction of sexual attraction. Two such structures are the *SDN of the preoptic anterior hypothalamus* and the *suprachiasmatic nucleus* (SCN) of the hypothalamus. Researchers Swaab, Gooren, and Hofman (1992) examined the hypothalamus of 34 subjects: 10 nondemented "gay" men who had died of AIDS, 6 nondemented "heterosexual" men and women (four male and two female) who had died of AIDS, and 18 subjects who died of other causes and whose sexual orientation could not be established. The latter group served as a reference group for comparison. The SDN of the hypothalamus is typically twice as large in men as in women (Swaab & Fliers, 1985). Although clear sex differences were evident, Swaab, Gooren, and Hofman (1992) found no differences in the size of the SDN of the preoptic anterior hypothalamus between "gay" and "heterosexual" men, contrary to what the androgen underexposure theory predicts. Dörner's (1988) contention that "gay" men have a female hypothalamus was not upheld. In fact, sex differences in the SDN of the anterior hypothalamus may have nothing to do with prenatal hormones. Swaab and Hofman (1988) argue that sex differences related to the SDN of the anterior hypothalamus are a result of postnatal events, not prenatal hormones. Sex differences in this area of the hypothalamus occur after age four in females, as a result of cell death and age (Hofman & Swaab, 1989). Furthermore, because sex differentiation in the SDN of the preoptic anterior hypothalamus takes place postnatally, psychosocial experiences are also likely to interact with brain development and age to contribute to this difference. Whatever events result in cell death in the SDN of the preoptic anterior hypothalamus in "heterosexual" women, "gay" men and "heterosexual" men appear to be protected. Their SDNs are similar in size.

Swaab, Gooren, and Hofman (1992) also discovered that the volume and cell count of the SCN, another hypothalamic nucleus, was twice as large in "gay" men as for the reference group, confirming an earlier finding by Swaab and Hofman (1990). Volume and cell counts of the SCN peak around age 13 to 16 months in newborns, then decline. Yet "gay" men possessed a

SCN similar in size to that of newborns. The size of the SCN of the reference group was at the expected adult level—35% of peak value. For some reason, programmed cell death in the SCN—most likely the result of postnatal experiences—failed to occur among "gay" men in the sample. Although interesting and statistically significant, just how the SCN influences the direction of sexual attraction is not clear. In mammals, the SCN regulates sleep and activity cycles and is not known to influence sexual behavior. There is no information available regarding the size of the SCN among "lesbians."

Likewise, the relationship between size of the anterior commissure and sexual orientation is difficult to understand. As noted earlier, women have a larger anterior commissure than do men (Allen & Gorski, 1991). In addition, Allen and Gorski (1992) reported that "gay" men have a larger anterior commissure than "heterosexual" men, and the anterior commissure among "gay" men is approximately the same size as in "heterosexual" women. Although at first this finding appears to support the androgen underexposure theory of sexual orientation, such an effect could also be a function of postnatal social experiences. The anterior commissure facilitates the transfer of visual information during cognitive tasks and, thus, would be influenced by experience (Hooper, 1992). Similar to the SCN, this region is not known to influence sexual attraction.

In sum, the androgen under- and overexposure theories of sexual orientation receive mixed support from neuroanatomic studies. Partial support is offset by contradictory findings and by alternative explanations that postnatal social experiences contribute to neuroanatomic sex differences.

4.1.4.1 Otoacoustic Emissions in the Auditory System

A relatively recent and intriguing area of study regarding sexual orientation and the androgen under- and overexposure theories is the anatomic structure of the human inner ear—more specifically, sounds emanating from the inner ear as a result of structural differences. A number of sex differences also exist in the auditory system, which Dennis McFadden (1998) described in a recent review. For example, external ear canals and middle-ear volumes are slightly larger for men than for women. Men are also better at auditory discrimination tasks than women, including detecting differences in sound time, intensity, and localization. Still, on the average, adult women are about 3 dB more sensitive than men for frequencies of 2000 Hz and above. Sex differences in hearing sensitivity are evident in childhood.

Yet one of the most interesting sex differences is related to sounds produced in the inner ear. Normal cochleas make sounds, as well as receive and process sounds. These sounds are called *otoacoustic emissions* (OAEs) and can be detected by a miniature microphone placed in the external ear canal. OAEs come in several forms, but we are concerned with only two: *spontaneous otoacoustic emissions* (SOAEs) and *click-evoked otoacoustic emissions* (CEOAEs). SOAEs refer to one or more tonal sounds that are continuously produced by most normal-hearing ears. SOAEs are not heard by the owner of the ears and are not the source of tinnitus. The frequency and number of SOAEs vary by ear and by sex (McFadden, Loehlin, & Pasanen, 1996). SOAEs are more frequent for right ears than left ears, and adult women typically experience more SOAEs than men. CEOAEs are echoes emanating from the fluid-filled cochlea into the air-filled middle and outer ear in response to a brief acoustic stimulus such as a click. The echo response is quick and weak. Usually hundreds of responses are averaged to illustrate them. On the average, CEOAEs are 2 dB to 3 dB stronger in females than in males (McFadden, Loehlin, & Pasanen, 1996). CEOAEs are also stronger for right ears than left. What is more, similar sex difference patterns in SOAEs and CEOAEs are evident in children and remain relatively stable (Burns, Arehart, & Campbell, 1992; McFadden, 1998).

Sex differences in OAEs are so distinct and persistent that researchers have hypothesized that prenatal androgen exposure is the cause (McFadden, 1998), and some data support this contention. Opposite-sex dizygotic females are more similar to their male co-twins than they are to other females (McFadden, Loehlin, & Pasanen, 1996). That is, dizygotic females with a male co-twin have fewer SOAEs, weaker CEOAEs, and reduced hearing sensitivity compared with other females. Prenatal androgen may have an organizational effect on the auditory system, as well as on other areas of the brain. However, the postnatal cochlea also responds to sex hormones. McFadden (1998) notes that hearing sensitivity varies for women during the monthly menstrual cycle, although the exact mechanism for this variance is not well understood. More precise measurement and methodology are needed to illuminate these findings. Furthermore, one study of two preoperative male transsexuals treated with estrogen suggests that androgenized cochleas respond to hormone changes (McFadden, Pasanen, & Callaway, 1998). One of the two male transsexuals who had not received estrogen prior to the study evidenced multiple SOAEs following estrogen treatment, although none had been evident previously.

McFadden and Pasanen (1998, 1999) measured otoacoustic emissions in 237 males and females who identified as either "heterosexual," "homosexual," or "bisexual" and who engaged in sexual fantasies or behavior consistent with their sexual identity. Contrary to what the androgen under- and overexposure theories predicts, no differences in otoacoustic emissions— SOAEs or CEOAEs—were found between "homosexual" and "heterosexual" males. Neither did "bisexual" males differ statistically from other males, although the number of "bisexual" males was small. In short, men of different sexual orientations have "masculinized" auditory systems; "gay" and "bisexual" men do *not* have "feminized" auditory systems. Because changes to the auditory system are thought to occur as a result of sex differentiation, the mechanism that accounts for sexual attraction in males is likely to occur independently and differently from sex differentiation.   McFadden and Pasanen (1998) suggest that variation in timing and concentration of androgen at different brain sites may better account for differences in male sexual orientation and a fully "masculinized" physiology.   I also suggest that the *ratio* of prenatal sex hormones may be important to sexual attraction.   I will return to this idea again in chapter 5.

Contrary to the data on men, McFadden and Pasanen (1998, 1999) found distinct variability among otoacoustic emissions between women of different sexual orientations.   "Homosexual" and "bisexual" women had similar SOAEs and CEOAEs that were intermediate to "heterosexual" women and men.   In other words, the OAEs of "homosexual" and "bisexual" women were more "masculine" or less "feminine" in effect.   McFadden and Pasanen suggest that prenatal exposure to higher than typical levels of androgens "masculinizes" the peripheral auditory system and the direction of sexual attraction among "lesbians" and "bisexuals."   Perhaps very low concentrations of androgens "feminize" sexual attraction and the auditory system for women during a similar critical period.   Physiology differentiates sexually during different critical periods.   McFadden and Pasanen's findings are the strongest evidence to date that same-sex female sexual attraction is a function of prenatal androgen exposure.   Studies of otoacoustic emissions among males and females in non-Western cultures by direction of sexual attraction and erotic fantasy would help support or disconfirm a possible biological substrate of sexual orientation.

These data also indicate that different processes underlie sexual orientation for males and females.   McFadden and Pasanen (1998, 1999) emphasize that development of brain structures occurs at different rates for males and females and that particular brain sites may be differentially sensitive to the timing and concentration of androgens.   Bailey, Gaulin, Agyei, and Gladue (1994) also point out that systems that differentiate sexually earlier or later are not subject to the same developmental influences. Consequently, a great deal of variability is possible in sex-typed features and direction of sexual attraction.   Critical periods for different emphases in sexual attraction and sex differentiation may overlap somewhat (McConaghy, 1993).   Perhaps for females critical periods for same-sex erotic predisposition and sex differentiation of the auditory system overlap, whereas for males critical periods for sex differentiation of the auditory system and same-sex erotic predisposition have no overlap.

4.1.5 Functional Differences

Structural differences in the brain are difficult to find, but sex differences in cognitive functioning are clear.   Differences in cognitive functioning are also evident across sexual orientations, although not always as predicted by the androgen under- and overexposure theories.   According to these androgen theories, "gay" men *should* exhibit cognitive abilities similar to "heterosexual" women, and "lesbians" *should* respond more like "heterosexual" men on cognitive tasks.   Some but not all predictions are supported by the data.

At rest, "gay" men exhibit different *electroencephalographic* (EEG) activity than "heterosexual" men and women (Alexander & Sufka, 1993). However, during affective judgment tasks for verbal and spatial stimuli, "gay" men show EEG patterns more similar to "heterosexual" women.   What is more, EEG patterns of "gay" men are "dynamic" and vary with "degree of homosexuality" (McKnight, 1997, p. 42).   The reason for these differences is not certain.

Another study employed a new technique for mapping activity in the brain, *magnetoencephalography* (MEG), which measures magnetic rather than electrical fields.   Magnetic readings of brain activity may be more accurate than conventional EEG because the brain is "electrically resistant but magnetically transparent," according to the study's author (Beatty, 1995).   In previous research with MEG, auditory stimulation evoked asymmetrical and lateral reactions between brain hemispheres among males, being more anterior in the region of the superior temporal *gyri* of the right hemisphere. Females, however, show a more symmetrical pattern of brain activity when stimulated by sound.   These findings suggest that males generally process auditory information in one hemisphere, and when using both hemispheres do not process in comparable locations in each hemisphere.   Females, instead, generally process auditory information in both hemispheres and in similar bilateral locales.   Traditionally, sex differences in cognitive functioning have been attributed to male exposure to prenatal androgens.   The androgen underexposure theory of sexual orientation predicts that "gay" men function cognitively more like "heterosexual" women.   Consistent with this prediction, eight "gay" men failed to evidence significant laterality or hemispheric asymmetry in response to auditory stimuli as measured by the MEG, unlike the nine "heterosexual" men (Reite, Sheeder, Richardson, & Teale, 1995). "Heterosexual" men demonstrated clear laterality and anterior reactions in the right hemisphere to auditory stimulation.   Unfortunately, females were not included in this study, so responses by "gay" men could not be compared. This study can conclude only that "gay" and "heterosexual" men do not process auditory stimuli similarly.   Further studies are needed to corroborate and clarify these findings.

Other researchers have directly compared "gay" men and "lesbians" with "heterosexual" men and women on established sexually dimorphic traits, such as visuospatial abilities and verbal skills. In general, men demonstrate greater visuospatial abilities, and women have superior verbal skills (Halpern, 1986). Consistent with this finding, women have proportionally larger Wernicke and Broca language areas than men (Harasty, Double, Halliday, Kril, & McRitchie, 1997). Gladue, Beatty, Larson, and Staton (1990) matched groups of "heterosexual" and "homosexual" men and women on age, education, occupation, and exposure to activities that promoted spatial skills and then assessed their subjects' visuospatial and verbal abilities and degree of "masculinity" and "femininity." Low masculinity scores are associated with poorer spatial abilities (Signorella & Jamison, 1986). Gladue and associates found that "gay" men performed less well than "heterosexual" men on two measures of visuospatial skills but performed similarly on other cognitive abilities. Masculinity scores and spatial abilities were not strongly related to sexual orientation for males. Curiously, however, "lesbians" performed more poorly than "heterosexual" women on only one measure of spatial ability (the water level test) and demonstrated no differences on other measures. Ratings of masculinity and femininity were nearly identical for both groups of women. Thus, somewhat consistent with the androgen underexposure theory of sexual orientation, "gay" men performed more like "heterosexual" women on visuospatial tasks but performed similarly to "heterosexual" men on verbal tasks. However, contrary to the androgen overexposure theory, on most measures, "lesbians" and "heterosexual" women responded similarly. There was no evidence of "masculinized" cognitive functioning among "lesbians."

In a similar test of cognitive abilities, "gay" men and "heterosexual" men and women were matched on handedness, education, and age (McCormick & Witelson, 1991). Handedness has been linked to variations in prenatal sex hormones and same-sex erotic attraction (Geschwind & Galaburda, 1985; McCormick, Witelson, & Kingstone, 1990; Nass et al., 1987; Netley & Rovet, 1982). Compared with the general population, "gay" men more often show a non-right hand preference. McCormick and Witelson (1991) compared matched samples of "gay" and "heterosexual" men against a control group of "heterosexual" women on measures of spatial ability and verbal fluency. Consistent with Gladue et al. (1990), McCormick and Witelson found that "gay" men performed more poorly on visuospatial tasks than did "heterosexual" men. In fact, "gay" men's scores were intermediate to responses by "heterosexual" men and women on three spatial tasks and on one fluency task. "Heterosexual" men and women demonstrated the expected sex differences in spatial skills relative to fluency abilities. "Gay" men, however, had *similar* spatial and fluency scores. The difference between scores for "gay" men and "heterosexual" women was not statistically significant. In other words, "gay" men did *not* evidence the male-typical pattern on spatial and fluency tasks. *Nor* did "gay" men evidence a female-typical pattern on these tasks.

McCormick and Witelson (1991) commented on the considerable within-group variability of handedness among their participants. Most functional differences between "gay" and "heterosexual" men were found among nonconsistent right-handers (nonCRH). Except for one visuospatial task (the water level test), consistent right-handed (CRH) "gay" men responded more like CRH "heterosexual" men. There were no differences between CRH and nonCRH "gay" men on verbal fluency tasks. However, CRH and nonCRH "heterosexual" men demonstrated more variability than "gay" men on fluency tasks. Handedness was also a factor for "heterosexual" women on one fluency test (digit symbol), with nonCRH women outperforming CRH women.

In sum, cognitive functioning data regarding sexual orientation suggest that "gay" men process information differently than "heterosexual" men and women. "Gay" men perform intermediate to "heterosexual" men and women, providing male-typical responses on verbal fluency tasks and more female-typical responses on visuospatial tasks in general. It is possible that the hormonal mechanisms that account for same-sex erotic predisposition in "gay" males may also influence cognitive organization and functioning. Prenatal exposure to androgens in males is generally associated with CRH, laterality in cognitive functioning, and high visuospatial-low verbal fluency performance. CRH "gay" and "heterosexual" men responded similarly on most visuospatial tasks, and CRH and nonCRH "gay" men performed similarly on verbal fluency tasks. NonCRH "heterosexual" men demonstrated a great deal of variability in their performance scores compared with CRH "heterosexual" men. It would be interesting to know the degree to which nonCRH "heterosexual" men have same-sex erotic fantasies or experiences compared with CRH "heterosexual" men. Unfortunately, neither of these studies assessed sexual orientation with that degree of precision. At any rate, this pattern of male cognitive performance by sexual orientation and handedness also supports the idea that the critical period for same-sex male sexual attraction precedes slightly and overlaps with critical periods for CRH and male-typical cognitive organization. Critical periods for other-sex erotic attraction in males may coincide with establishment of CRH and male-typical cognitive organization. For males, current data regarding cognitive functioning are consistent with the androgen underexposure theory of sexual attraction.

However, the data also suggest that different mechanisms affect male and female sexual orientation. For example, the pattern of cognitive performance for "lesbians" is not consistent with the androgen theory (Gladue *et al.*, 1990). Although the data are very limited, "lesbians" appear to have cognitive abilities that are similar to "heterosexual" women. In the one study discussed here that compared "heterosexual" women and "lesbians," test performance differed only on one visuospatial measure, with "lesbians" women performing more poorly. "Lesbians" women did not evidence male-typical performance patterns. In their review of this literature, Collaer and Hines (1995) concluded that early exposure to high levels of androgens influences cognitive abilities in genetic females, resulting in more male-typical performance patterns, whereas excessive exposure to estrogen has little apparent affect on cognitive functioning. However, most genetic females exposed to excessive amounts of androgens or estrogens identify and behave as "heterosexual." Thus, the relationship between cognitive functioning, sexual orientation, and prenatal hormone exposure is more complicated for women. Additional studies comparing women on sex-typed cognitive tasks and on handedness are necessary before any conclusions can be drawn regarding the timing and organizational effects of prenatal hormones on female sexual attraction.

4.1.6 Summary of Androgen Theory Research

Across the five main areas of research based on androgen under- and overexposure theories, clear and consistent support for these theories has not emerged. Only studies of otoacoustic emissions among women and cognitive functional differences among men have demonstrated results consistent with these theories.

The obvious explanation for the lack of confirming data for the androgen under- and overexposure theories is that they are bad. Indeed, what I intend to demonstrate in the remainder of this chapter and in the next one is that these androgen theories are too simplistic. A more comprehensive explanation of sexual attraction emphasizes the ratio and timing of prenatal sex hormones and the interaction between sensitized neural networks and social experiences.

4.2 OTHER BIOLOGICAL RESEARCH IN SEXUAL ORIENTATION DEVELOPMENT

Other promising biological studies of sexual orientation give less attention to the androgen theories. These studies focus on non-androgen mechanisms of action, interactive processes, genetic factors, or nonerotic heritable traits related to evolved mating strategies. Although concrete conclusions from this line of research are few, these studies provide additional circumstantial support for a biological substrate of sexual attraction.

4.2.1 Family Prevalence and Genetic Studies

One means of determining the *heritability* of a particular trait is to demonstrate that the trait is highly probable among genetically similar individuals—which is *not* to say that the trait is inherited. (I discuss the transmission of nongenetic traits later.) Identical or *monozygotic* (MZ) twins share all of their genes. Therefore, MZ twins inherit the same genetic propensities. MZ twins also share many nongenetic traits, given their highly similar development and social environment. However, if the direction of sexual attraction is inherited, MZ twins should often have similar sexual orientations. On the other hand, fraternal or *dizygotic* (DZ) twins share only half of their genes. Ordinary siblings share only 25% of their genetic material, and individuals from different families do not share genes. Actually, the relative genetic difference between unrelated individuals is quite small but significant. Only about 1% of genetic material differs between unrelated individuals (Hamer & Copeland, 1994). However, that difference amounts to roughly *3 million base pairs of DNA* out of 3 billion.

In short, highly heritable traits should be common among genetically similar individuals. Between individuals who have little or no genes in common, the trait in question should occur no greater than chance. Although highly shared environments can increase the occurrence of non-inherited traits among differently related individuals, environments are similar but never identical. Even among MZ twins, social environments and interpersonal experiences are never identical and not always similar. The relationship between genetic relatedness among sibling pairs and trait prevalence is called *concordance*, which is expressed as a proportion from 0%—meaning, no genetic basis—to 100%—meaning, an inherited genetic trait.

Two phenomena that are strongly heritable are schizophrenia and diabetes mellitus. Schizophrenia, for example, has concordance rates of 45% percent for MZ twins and 15% for DZ twins (Mange & Mange, 1990). Diabetes mellitus Types I and II are 50% and 95% concordant for MZ twins and 5% and 25% concordant for DZ twins, respectively. Clearly, Type II diabetes is more likely to be inherited than Type I. If sexual orientation is also a heritable trait, then the same form of sexual attraction should appear more frequently among genetically similar individuals, such as MZ twins, than among genetically dissimilar individuals.

One of the earliest twin concordance studies of same-sex male sexual attraction was conducted by Franz Kallmann and published in 1952. Surprisingly, Kallmann found 100% concordance for same-sex erotic attraction among 40 male MZ twins! It appeared that same-sex erotic attraction among men was definitely a heritable trait. Not surprisingly, however, other researchers failed to replicate Kallmann's findings. In fact, several significant methodological errors and inconsistencies in Kallmann's study account for his unusual results. Describing Kallmann's research design flaws also illustrates how unexamined cultural assumptions influence the practice of science. First, most of Kallmann's subjects came from prisons and mental institutions. Second, he was vague about his sampling methods, which suggests nonrandom selection. Kallmann acknowledged that he did not interview all of the DZ co-twins to check consistency and reliability, but he implied that all of the MZ co-twins were interviewed which seems very unlikely. Third, Kallmann provided no information about how he determined the sexual orientation of his subjects. As we have discussed already, sexual orientation is not so easy to determine. It is possible that he asked his subjects about their erotic feelings, fantasies, and sexual behavior and then looked for consistencies in self-report. However, because he gave no description, it is likely that he subjectively classified his subjects. Fourth, and most significantly, Kallmann noted in his article that his research was motivated by the belief that "adult homosexuality continues to be an inexhaustible source of unhappiness, discontent, and a distorted sense of human values" (1952, p. 296). Here are his bias and his objective. Kallmann believed that same-sex erotic attraction produced unhappiness and discontent and so looked for subjects among the extremely unhappy and discontent—criminals and mentally disturbed patients. It probably never occurred to him to examine same-sex erotic attraction among ordinary well-functioning folk.

No researcher has come close to replicating Kallmann's findings, which in itself does not deny a biological basis for sexual orientation. In fact, recent larger, more methodologically sound studies have found that the prevalence of same-sex erotic attraction among twins is consistent with a biological trait. In one study, Bailey and Pillard (1991) compared 115 male co-twins, where one twin was "gay." Fifty-two percent of 56 male MZ twins were concordant for same-sex erotic attraction. Twenty-four percent of 54 DZ twins were also concordant for same-sex attraction. Nontwin biologic brothers and unrelated adoptive brothers of a "gay" twin were 9% and 11% concordant, respectively, for same-sex attraction. Bailey and Pillard concluded that these concordance rates are exactly what would be expected for a highly heritable trait that is not completely governed by biology. Later, Bailey and Pillard (1993) noted that MZ twins in their study who were both "gay" were much more likely to have "gay" nontwin brothers than were MZ erotic-discordant twins. Thus, consistent with the idea of a heritable trait, erotic concordance among MZ twins was associated with a higher incidence of same-sex erotic attraction among nontwin brothers. In addition, Bailey and Pillard also noted that two pairs of MZ twins in their original sample were separated at birth and reared apart, reducing the probability of highly similar social environments and interpersonal experiences. All of the MZ co-twins who were reared apart identified as "gay."

Another group of researchers, Whitam, Diamond, and Martin (1993), reported concordance rates similar to Bailey and Pillard for same-sex male sexual attraction. Subjects in this study were solicited through public advertisements. Sixty-five percent of 34 MZ co-twins and 29% of 14 DZ co-twins were concordant for same-sex erotic attraction, where one twin was "gay." This pattern of declining concordance is consistent with a genetically linked trait. The Whitam *et al.* (1993) study also included three sets of triplets—extremely rare finds and very important for concordance studies. Among the first set of triplets, the two MZ brothers were "gay" and their DZ sister was "heterosexual." Among the second set of triplet sisters, two MZ co-twins were "lesbian" and their DZ sister was "heterosexual." And among the third set of triplets, all three MZ brothers were "gay" and shared remarkably similar sexual histories. What is more, two sets of male MZ co-twins in this sample were reared apart. One set of MZ co-twins was concordant for same-sex erotic attraction; the other set was discordant. Taken together, these two studies provide supportive circumstantial evidence—especially among MZ co-twins reared apart—that same-sex erotic attraction is a heritable trait among males.

These studies also suggest that different influences affect erotic discordance among MZ twins. Whatever mechanisms determine sexual attraction, these mechanisms are similar but not identical even for MZ twins. Although these concordance studies were not intended to disprove androgen theory, nonetheless, the androgen underexposure theory cannot account for erotic-discordant MZ co-twins. This simply should not happen.

Bailey and Pillard (1993) also investigated concordance rates of same-sex female sexual attraction and found similar but less strongly heritable patterns for women compared to men. Study authors compared 71 MZ and 37 DZ co-twins, where one twin identified as "lesbian." In this sample, 48% of female MZ co-twins and 16% of DZ co-twins were concordant for same-sex erotic attraction. Nontwin biological sisters and adoptive sisters had concordance rates of 14% and 6%, respectively, for same-sex attraction or attraction to both sexes. Furthermore, "lesbian" twins were more likely to have "lesbian" sisters than "gay" brothers, a pattern that also mirrors the prevalence of same-sex eroticism among "gay" male siblings.

The concordance data concerning "gay" and "lesbian" twins (Bailey & Pillard, 1991, 1993; Whitam *et al.*, 1993) support three probable conclusions. First, same-sex erotic attraction is moderately heritable, meaning that this trait often but not always is evident among genetically similar individuals where one co-twin has the trait. These data support the contention that same-sex erotic attraction is a biological trait, although alternative social explanations cannot be ruled out. Traits can be transmitted socially and have no biological foundation. However, evidence that MZ co-twins and triplets who were raised apart identify as "gay," when one co-twin is "gay," tends to refute an absolute social-learning explanation. Similar environments are not identical, and same-sex eroticism is not a common occurrence. Second, same-sex erotic attraction is more heritable for males than females. The concordance data are stronger for males, which also suggests that different mechanisms account for same-sex eroticism in males and females. And third, the presence of a heritable trait does not result inevitably in development of the trait. Concordance rates for MZ twins are far from absolute. Given that MZ twins share the same genetic material, whatever biological mechanisms influence sexual attraction, they are not sufficient alone to determine same-sex erotic orientation.

Again, concordance data provide supportive circumstantial evidence but are also subject to alternate interpretations. McGuire (1995), a strong critic of concordance studies, complains that nonrandom sampling is an inherent flaw in such research and that other explanations for the trait are not ruled out. This is an overly harsh and uncharitable criticism. Twins are uncommon, and finding a sufficient number of twins where at least one co-twin is "gay" is very rare. To draw a random sample, researchers would need large numbers of this highly select sample. The probability of this event is so low that researchers who study twins and same-sex eroticism are unlikely to have anything other than nonrandom samples. True, twin studies, random or nonrandom, are correlational in nature and can only say that certain traits are likely to occur together. Although correlational studies can never rule out alternate explanations for results, some explanations are less probable than others. (Recall also the early admonition that Science is based on probability, not certainty). Highly similar social environments, especially among genetically similar individuals, may increase the likelihood of shared traits that are not biological. However, as Bailey, Pillard, Neale, and Agyei (1993) point out, even MZ twins do not experience identical environments or have the same traits, although they have the same genes. Yet given the intense historical abhorrence of homosexuality in Western culture, it is not only improbable but also very unlikely that parents or family members *unintentionally* or *consistently* foster same-sex erotic attraction in any of their children. As noted in chapter 2, accidental social experiences that occur frequently enough to reinforce same-sex erotic attraction in a culture that punishes same-sex erotic attraction and highly rewards other-sex attraction are unlikely. Although concordance studies are not intended to provide causal data, they do, however, lend support to the contention that same-sex erotic attraction has a biological foundation.

It is worth responding to the credibility of criticism directed at concordance studies because genetic studies of sexual orientation, also correlational, have received similar criticism. One of the most publicized studies of sexual orientation in recent years involved the correlation of genetic markers on the long arm of the X chromosome among "gay" brothers. In 1993, geneticists Dean Hamer, Stella Hu, Victoria Magnuson, Nan Hu, and Angela Pattatucci published their findings in the journal *Science*. *Gene markers* are large segments of genetic material. Markers suggest the presence of one or more genes that control the expression of a trait (Hamer & Copeland, 1994). *Genes* are chains of protein sequences located close together. Genes are not discrete segments of matter.

Hamer and associates did *not* identify genes that cause same-sex erotic attraction. They identified common markers for male same-sex eroticism among related individuals with the same trait. To enhance their chances of finding positive results, Hamer and associates limited expression of the trait to maternal transmission. Therefore, genes "for" the trait would be specific to the X chromosome. Consequently, "gay" men with "gay" fathers—who might transmit the trait on the Y chromosome—were excluded from the study. In their sample of 114 "gay" men, 40 men had "gay" brothers. The 40 pairs of brothers were compared on 22 different genetic markers located on the tip of the X chromosome—an area known as Xq28. More than 200 genes exist between these marker boundaries. Amazingly, 33 of the 40 pairs of brothers (83%) were concordant for a series of five genetic markers. Study researchers concluded that these five markers play "some role in about 5 to 30% of gay men" for whom same-sex erotic attraction is inherited through the X chromosome (Hamer & Copeland, 1994, p. 146). In other words, Hamer and colleagues believe that one or more genes located between these five markers predispose this form of maternally transmitted same-sex male erotic attraction.

To test their hypothesis, Hamer and Copeland (1994) predicted that the "gay" men in their study *should* have more brothers and maternal uncles who are "gay" than they have "gay" paternal relatives because of their common X chromosome link. Indeed, this is what they found. Among the 40 pairs of "gay" brothers, seven of 96 maternal uncles (7.3%) were also "gay" compared to only two of 119 paternal uncles (1.7%). Furthermore, since sisters share up to 75% of their genes, the proportion of "gay" maternal cousins through the mother's sister *should* also be high for the sample of "gay" brothers. It was. Four of 52 maternal cousins (7.7%) through a maternal aunt were "gay," as opposed to only two of 51 maternal cousins through an uncle (3.9%).

Next, Hamer and colleagues replicated their original findings with a second group of 33 pairs of "gay" brothers (Hu *et al.*, 1995). Consistent with the first study, 67% of "gay" brothers shared the series of five genetic markers at Xq28. Although 67% is considerably lower than the 83% concordance rate reported in the original study, it is still well above chance probability. Hamer and colleagues also noted that the five markers showed an inverse correlation (negative 22%) between "gay" men and their "heterosexual" brothers. That is, "heterosexual" men shared few of the five markers with their "gay" brothers. In other words, these markers are not related to other-sex erotic attraction in males. The locale of genes associated with other-sex erotic attraction in males is unknown and to date uninvestigated. In addition, Hamer and colleagues found no correlation between "gay" men and their "lesbian" sisters for Xq28 markers (Hu *et al.*, 1995). The genes that contribute to same-sex eroticism in men are not the ones that contribute to same-sex eroticism in women.

On the basis of these two genetic linkage studies, Hamer and Copeland (1994) concluded that one form of maternally transmitted same-sex erotic attraction in males is a *polygenetic* X-linked trait located on Xq28.

Genes in other locations probably regulate different forms of male-male sexual attraction. In addition, genes linked to maternal transmission of same-sex eroticism among males are not the ones that influence same-sex female eroticism. Genes associated with same-sex female attraction are located someplace other than Xq28 (Pattatucci & Hamer, 1995). To date no published studies have investigated similarities between genetic markers for "gay" brothers and their "heterosexual" sisters or for "lesbians" and "heterosexual" men. However, it is also entirely possible that genes on Xq28 are not specific to same-sex eroticism and that genes in other locations more directly influence sexual attraction. Genes on Xq28 may indirectly influence the development of same-sex attraction in males through as yet unidentified and unmeasured traits.

Genetic linkage studies are particularly complicated, technical, and vulnerable to confusion and misinterpretation because such tiny bits of material are said to influence such major traits as sexual attraction. Consequently, disagreements among scientists about these analyses are many. One set of critics, Risch, Squires-Wheeler, and Keats (1993), complained that Hamer's data is questionable because he examined nonrandom subjects obtained through advertisements—a common method of recruiting twins. Risch and colleagues also complained that Hamer misapplied Mendelian processes to complex traits, used inappropriate inferential population estimates, and overinterpreted the published twin data. Much of their complicated argument has to do with calculating the most appropriate LOD (logarithm of the odds ratio) score for comparing expected genetic linkage. Risch and colleagues predicted an inverse proportional relationship for same-sex attraction among maternal relatives compared with population prevalence. Specifically, they argue, the prevalence of same-sex erotic attraction should be higher for uncles (second-degree relatives) than for cousins (third-degree relatives), contrary to what Hamer and colleagues reported. Risch and associates (1993) went on to state that "these results are not consistent with any genetic model" (p. 2064). Hamer et al. (1993) responded that Risch's inferential estimates of trait recurrence are only true if the two comparison proportions are measured on the same population, which was not the case in the Hamer studies. "Gay" brothers who were screened for the absence of direct father-to-son transmission of the trait represented a select subsample of all "gay" men. Furthermore, Hamer and colleagues argued that Risch's predictions of trait prevalence were based on autosomal dominant traits and "not for an X-linked trait for which the expected order of rates in cousins and uncles is reversed" (1993, p. 2065). Contrary to Risch's statistical assumptions, Hamer and associates asserted that their study was designed only to identify a possible correlation between a particular chromosome region and a trait, not to estimate the role of this region in the population at large. Finally, they noted that they applied a LOD threshold score of four, which is 10 times higher than the conventional level of three, in order to reduce the possibility of a false positive Type I error in analyses with a complex nonMendelian trait.

Only exact replication studies with uniform methodologies will resolve differences in interpretation of linkage data. Except Hamer's own replication study (Hu et al., 1995), just one other genetic linkage study has been reported. In a self-described replication of Hamer's original study, Rice, Anderson, Risch, and Ebers (1999) recruited 182 "gay" men through advertisements in gay newsmagazines. "Gay" men with "gay" fathers were not excluded, as in the Hamer studies. The 182 men had 614 brothers, of whom 269 (44%) were also "gay." Among their 270 sisters, 49 (18%) were "lesbian." Rice and colleagues reported DNA analyses on 52 pairs of "gay" brothers for four genetic markers on Xq28. However, only two of the four markers appear to be the same ones assessed by Hamer and colleagues (1993; Hu et al., 1995). Not only did Rice and colleagues not measure the same kind of individuals, they did not assess the same number of markers nor even the same markers as Hamer. Despite their claim, the Rice et al. (1999) study is not a replication of Hamer's work. At any rate, Rice and colleagues reported a genetic linkage among "gay" brothers of 43% for the four Xq28 markers, which is below the proportion expected by chance. Study authors also misleadingly combine their replication study with Hamer's data to illustrate that the genetic sharing between "gay" brothers is 55% at best, which is at the chance level. Rice et al. (1999) demonstrated that pairs of "gay" brothers do not share four Xq28 markers above chance levels. However, a true replication of Hamer's work has yet to be done.

4.2.1.1 Summary of Family and Genetic Studies

Concordance and genetic linkage studies of "gay" sibling pairs and their families provide supportive circumstantial evidence that same-sex erotic attraction in males has a biological basis and is heritable. Evidence that MZ co-twins, especially those reared apart, are likely to correspond for same-sex eroticism provides the strongest support for a biological substrate. Far fewer linkage studies have included women, although sib pair and family studies suggest a somewhat weaker biological basis for same-sex attraction in females. Whatever biological mechanisms influence same-sex eroticism, these mechanisms differ for males and females.

Concordance and genetic linkage studies are correlational in nature, and do not suggest causality. Hamer's work has yet to be replicated and fully understood, and concordance studies demonstrate that biological factors are one of many that contribute to same-sex erotic attraction. In chapter 5, I hypothesize that sex-specific erotic attraction is biologically predisposed and fully dependent upon the social environment and personal experience for expression, consistent with distinct sex differences. This new model makes predictions that are similar to findings from concordance and family studies.

4.2.2 Estrogen Studies

By assigning androgen the definitive role in sexual development, conventional biological researchers have minimized and in most cases ignored the role of estrogen. The conventional assumption is that estrogens make no or little significant contribution to sex differentiation or to sexual orientation. Researchers are unclear exactly how estrogens contribute to sexual development, because their effects are seldom studied. However, during prenatal development, mammals are exposed to significant quantities of estrogen—some exogenous hormones from mother and some endogenous hormones in differing ratios depending on the genetic sex of the fetus and the absence of hormonal anomalies. All mammals produce androgens and estrogens, proportionate to their sex (Hutchison & Beyer, 1994). Levels of circulating testosterone in men are about 10 times higher than for women, while estrogen levels are around 10 times higher in women.

It is improbable that estrogen has no effect on sexual development or on sexual attraction and that androgens alone turn "on" or "off" different kinds of eroticism (Harding, 1986). We have already seen that the androgen under- and overexposure theories fail to explain the development and variety of human sexual attraction. In fact, high levels of prenatal estrogen may be important to healthy fetal development. Estrogen may moderate the influence of testosterone by competing for the same receptor sites or modifying receptor sensitivity. And the appropriate ratio of both hormones may be critical for healthy sexual functioning.

One group of researchers investigated the role of estrogen in the sexual behavior of mice by genetically engineering a strain of male mice that lacked estrogen receptors (ER) but retained androgen receptors (Ogawa, Lubahn, Kurach, & Pfaff, 1997). Male mice were exposed to typical levels of maternal estrogen and produced their own estrogen, to no effect. Ogawa and colleagues found that adult ER-deficient male mice exhibited less efficient sexual behavior—although sexual motivation was unchanged—compared with unaltered mice. Aggression was also diminished among ER-deficient male mice, although testosterone levels were usually high. In unaltered mice, estrogen receptors provide negative feedback about androgen levels. However, in the absence of estrogen receptors to provide that feedback, testosterone production was unregulated. Yet high levels of testosterone did *not* increase sexual activity or aggression in ER-deficient male mice. In mice, male-typical behavior may require the presence of both sex hormones.

A similar effect was demonstrated in ER-deficient female mice. Without estrogen receptors, female mice also produced greater than typical quantities of androgen. In unaltered female mice, high levels of circulating androgens typically increase the frequency of sexual behavior. However, ER-deficient female mice with high levels of circulating androgens *never* engaged in lordosis—sexually presenting behavior (Ogawa *et al.*, 1997). Apparently, female mice also need both sex hormones and hormonal communication through feedback mechanisms in order to engage in functional sexual behavior. The next critical step in this line of research is to investigate the relationship between sex hormones and sexual behavior in other species, particularly primates and humans. It is crucial to know whether estrogen moderates or facilitates the effect of androgen among primates.

Estrogens and androgens are known to compete for some of the same receptors (Harding, 1986). Steroidal and nonsteroidal estrogens, for instance, can occupy androgen receptor sites, and androgens such as 5 alpha- and 5 beta-dihydrotestosterone can occupy estrogen sites. Competition among sex hormones for receptors at multiple brain sites may promote behavioral variability and influence a number of complex social behaviors such as communication, personality, libido, and the direction of sexual attraction. Among male zebra finches, for example, the presence of sex hormones is required at eight different brain nuclei, as well as at the vocal organ, to produce song. Some brain sites in the zebra finch are receptive to one or both hormones, whereas the vocal organ is responsive only to androgen. Thus, for male zebra finches efficient singing involves the prior complex activation of multiple receptor sites by the appropriate sex hormone, given competition for those sites. It is doubtful that human sexual orientation is any less complex a behavior than is singing by zebra finches. Among mammals, peripheral tissue is typically sensitive to one sex hormone, whereas different brain tissues are receptive to one or both (Harding, 1986). Receptor sensitivity to hormones can be influenced by a host of factors, including age, circadian fluctuations, presence of other hormones, and relevant social stimuli.

As is the case for zebra finches, receptor activation by sex hormones creates the potential for *particular* kinds of learning. Well-timed subsequent social experiences are necessary to exploit the potential created by receptor activation. Endocrinologist Donald Pfaff (1980) explains how sex hormones and social experiences together produce efficient sexual behavior in rodents. Even in rats, sexual behavior is not a simple reflexive response to a single stimulus. Among adult female rats, efficient sexual behavior depends upon several early developmental events—exposure to appropriate perinatal sex hormones and particular environmental experiences—as well as a number of current interacting factors. Pfaff notes that for adult female rats sexual behavior is a chain of responses that creates a behavioral and neural *predisposition* for sexually presenting behavior *only after* physical contact on the flanks by a male rat. According to Pfaff, only if all the elements are in place is sexual behavior likely to occur. For male rats, the process of developing efficient sexual behavior is equally complex and dependent on early experiences. Besides perinatal exposure to androgen, efficient adult sexual behavior commences with male pups experiencing frequent anal-genital grooming by mother (Moore, 1986). Frequent anal-genital self-grooming usually continues after puberty. It is thought that genital licking increases testosterone production in male rats (LeVay, 1993). Male rats that were prohibited as pups from experiencing genital grooming by their mothers did not benefit from later social events and were not able to produce efficient sexual behavior. In short, *at least for rats, early hormonal events create the opportunity to benefit from particular social experiences that then permit certain behavior in the future, given the proper stimulus cues.* Somehow, this cumulative chain of events helps male rats develop fully as males and facilitates the detection of female receptivity, as well as mounting and penile-thrusting.

Among primates, efficient adult sexual behavior depends upon proper hormonalization, sufficient learning experiences, cues signaling receptivity, and the individual's social status relative to others in the troop. In studies of rhesus monkeys in open settings, several factors influence adult female sexual behavior, including hormone levels, the number of females and males present, and the individual's social rank (de Waal, 1984; Wallen, 1990). Rhesus females maintain a rigid social hierarchy. When multiple males are present, female competition is high, and rhesus females aggressively limit the frequency of sexual initiation by other females to times of highest sexual motivation—when level of circulating estrogen is high—unless the individual in question has high social rank. In that case, the alpha female initiates sexual activity with males without reprisals from the troop. A similar pattern of female troop-regulated sexual initiation has been observed among female gorillas and orangutans (Wallen, 1990). Even at times of highest sexual motivation, social comparison and behavioral interaction with other females moderate sexual responses by female primates.

Evidence linking sexual motivation among human females to hormone levels is mixed and weak, suggesting that sexual activity for women is motivated more by psychological and social factors than by biological ones. Nathaniel McConaghy (1993) attributes contradictory data regarding human female sex hormone levels and sexual motivation to several problems—difficulty in conceptualizing women's sexual behavior, inaccurately identifying phases of the menstrual cycle, poor measurement and recording of sexual behavior, and avoidance of sex by couples during menstruation for psychological reasons. Avoidance of sex during menses results in high frequencies of sexual activity on either side of menstruation, which masks sexual motivation during menstruation. Although McConaghy does not deny that circulating sex hormones motivate female sexual interest, he concludes based on the weak effects in hormonal studies that hormone levels have only a mild to moderate effect for women on sexual decision making. This view is consistent with predictions by evolutionary theory on female mating strategies and sexual motivation, an issue taken up later in this chapter.

Male fetal exposure to large amounts of estrogens in the form of *progestins* and the synthetic estrogen *diethylstilbestrol* (DES) has no apparent effect on sexual motivation or sexual orientation (Ehrhardt & Meyer-Bahlburg, 1981; Kester, Green, Finch, & Williams, 1980). However, Reinisch and Sanders (1992) detected evidence of "demasculinization" on neuropsychological testing among males exposed to DES during gestation. Unfortunately, the study authors did not assess sexual orientation. Testing evaluated hemispheric specialization of nonlinguistic spatial information and cognitive abilities, and in general men show strong spatial skills and hemispheric laterality, meaning that they show a hemispheric preference during processing of particular stimuli. Because aromatized estradiol—not synthetic estrogen—"masculinizes" the brain, there *should* be no cognitive differences between males exposed to DES and their unexposed brothers. Yet DES-exposed men demonstrated significantly less brain lateralization than did their brothers. About 20% of DES-exposed males showed a lateralized response compared to 80% of their brothers. DES-exposed males also demonstrated significantly less spatial ability than their brothers. Neither group differed on measures of intelligence. These data led Reinisch and Sanders to speculate that DES "feminized" and "demasculinized" boys, although unconverted and synthetic estrogens are not known to work in this way. An alternative explanation is that excess synthetic estrogen competed with androgens or aromatized estrogen for receptor sites in DES boys, diluting the "masculinizing" effects of androgens. Regardless, "feminization" or "demasculinization" of the brain as a result of DES exposure has no apparent effect on male sexual orientation (Ehrhardt & Meyer-Bahlburg, 1981; Kester *et al.*, 1980).

On the other hand, adult women who were exposed to progesterone-related compounds or DES *in utero* report an increased likelihood of same-sex erotic attraction and behavior (Ehrhardt, Meyer-Bahlburg, Feldman, & Ince, 1984; Ehrhardt *et al.*, 1985). Even so, most DES-exposed women are "heterosexual." Although excessive estrogen in females may function like androgen and promote so-called "masculine" effects such as same-sex eroticism, high levels of estrogen do not appear to set same-sex erotic attraction in women. At most, some females exposed to excess estrogen *in utero* have a predisposition toward same-sex erotic attraction, but many do not.

As a final note, the few estrogen studies to date have also dispelled the idea that prenatal estrogens "organize" female brain functioning. Conventional biological thinking has held that prenatal sex hormones, especially androgens, "organize" the brain in relatively fixed and permanent ways (Money, 1988). However, recent research has challenged the notion of fixed organizational and activational effects in female brain functioning (Rodriguez-Sierra, 1986; C. Williams, 1986). Among female rats, for example, brain organization and activation appear to occur at the same time. At puberty, high quantities of estrogen alter the morphology of the vagina and central nervous system in female rats, creating new neural pathways and synapses, *despite* prenatal organization. Simultaneous brain organization and activation at puberty may be a function of different receptor sensitivities or variable maturation rates of select groups of neurons (C. Williams, 1986). It may be possible that for female rodents estrogen receptors do not lose their plasticity when they mature (Rodriguez-Sierra, 1986). And not just nuclei, the two hemispheres of the brain mature at different rates (Nordeen & Yahr, 1983). Thus, for rats, a *quasiorganizational* effect occurs at various critical periods during the female's lifetime. With female neural plasticity comes the potential for greater capacity to evaluate and respond to changing social circumstance. Male rats are not known to exhibit this kind of neural plasticity in brain functioning. To date, no other published studies have reported simultaneous organizational and activational effects among females in other species. Whether human females possess neural flexibility in brain functioning is not known.

4.2.2.1 Summary of Estrogen Studies

For rodents and some non-human primates, estrogen plays a significant role in sexual motivation and in the efficient execution of sexual behaviors for both males and females. The extent to which these findings apply to human sexual attraction is unclear. However, animal data support the contention that efficient adult sexual behavior is dependent upon the appropriate combination of early hormonal events, a series of social experiences, and learning history. This cumulative interactive process mirrors the new model of human sexual attraction described in chapter 5.

Estrogen research on sexual attraction among humans is a largely unexplored territory. Knowledge that estrogen competes with androgen for particular receptor sites in the brain suggests that brain-based behaviors such as sexual attraction *should* be varied and *should not* be linked with gender.

4.2.3 Human Pheromones

Another intriguing area of research that has implications for understanding sexual attraction is work on human pheromones. *Pheromones* are neurochemical substances that signal sexual availability and readiness— usually through odor detection of hormones in urine or other secretions (Stern & McClintock, 1998). Primates and many non-human animals are known to excrete and respond to pheromones (Cutler, 1999). Humans may also respond behaviorally to pheromone signals in sweat or saliva. Winnifred Cutler, owner and president of Athena Institute for Women's Wellness Research, claims to have found human pheromones and markets a synthetic human male attractant pheromone for women. In spite of the financial motivation for her claim, some studies support the existence of human pheromones and their role in mate-selection strategies.

An early study of human pheromones attempted to learn why women who live together in close quarters, such as in dormitory residence halls, synchronize menstrual cycles (McClintock, 1971, 1984). The supposition is that women communicate neurochemically to each other the phase of their menstrual cycle through sweat, odor, and touch. Recently, Stern and McClintock (1998) found that physical contact with odorless compounds derived from women's armpit sweat during different phases of their menstrual cycle either accelerate or shorten the menstrual cycle of women. McClintock concluded that women have the potential to communicate pheromonally and speculated that, as in other species, humans possess other types of pheromones that are unrelated to ovarian function. Pheromonal communication between women may function to dampen reproductive competition and regulate reproductive potential among close groups of women—to sort of level the playing field. Within a close group, females who ovulate early would have a selective advantage over females who ovulate late.

A study by Wedekind, Seebeck, Bettens, and Paepke (1995) also found that women responded more positively and strongly to sweat odor of males who were immunologically dissimilar from them. Here, researchers assessed the *major histocompatibility complex* (MHC)—which plays an important role in immune function and creates proteins unique to self—in six college men. MHC is thought to help our bodies distinguish between self and other, and MHC is known to influence mate choice in mice. Women rated the odor of T-shirts worn for two nights by college men, and consistent with predictions women rated sweat odor from men who were MHC-dissimilar from them as more pleasant. However, women who were on oral contraceptives responded in just the reverse. Women on the pill rated odors from MHC-similar men as more pleasant. Wedekind and colleagues concluded that humans communicate to each other about reproductive viability through neurochemicals and speculated that female MHC detection aids in selection of genetically dissimilar mates with virile immune systems. What is more, oral contraceptives disrupt this communication and may even interfere with reproductively efficient mate choice.

Cutler (1999) conducted a double-blind, placebo-controlled study in which women were exposed to nonodorous axillary excretions collected from women across the menstrual cycle. Excretions were frozen, thawed, dissolved in ethanol, and applied to the upper lip three times per week for 14 weeks. Compared with the placebo group, women treated with pheromonal excretions reported a significant increase in sexual activity. Cutler,

Friedmann, and McCoy (1998) next tested a synthesized human male pheromone with 38 "heterosexual" men and found similar results. Cutler's description, however, does not make it clear whether the pheromonal substance for men was derived from female sweat, male sweat, or pooled sweat from both sexes. At any rate, men used the extract for six weeks. Compared with the placebo group, a significant number of men who used the pheromone extract reported an increased frequency of sexual intercourse and sleeping next to a romantic partner. The number of men who dated or self-masturbated was not statistically different. Cutler believed that sexual motivation is unaffected by the extract. Rather, she suggested that the partners of men who used the extract perceived them as more sexually desirable. This supposition has not been tested empirically.

Taken together, these studies suggest that both men and women detect and respond sexually to human pheromonal signals. Pheromonal communication is likely to aid mate selection and other reproductive strategies. When in close groups or in contact with female excretions, women synchronize their menstrual cycles and increase their frequency of sexual activity. Perhaps the detection of other proximate fertile women stimulates libido and sexual readiness. Similarly, men increase their frequency of sexual activity when in contact with synthetic pheromones, perhaps because they are perceived as more sexually desirable. Future human pheromone research should also investigate the effect among "gay" men and "lesbians." It seems unlikely that "gay" men or "lesbians" would respond favorably to other-sex pheromones.

In sum, research on human pheromones is in its infancy. However, limited data support the idea that sexual attraction has a biological component. Indeed, pheromone studies suggest that humans communicate neurochemically about reproductive viability and that this communication influences sexual behavior. Pheromone studies also nicely illustrate that the social environment influences biological functioning and sexuality.

4.2.4 Evolutionary Theories and Personality

Genes may only indirectly influence the direction of sexual attraction (McConaghy, 1993). Sexual orientation could reasonably be an accidental by-product of non-sexual genetic traits such as personality, or it could be an unintended condition of having a large brain. Or genes may not be involved at all, and sexual orientation could be passed along socially from one generation to the next (Scher, 1999).

Whether employing social evolution or natural selection, explaining other-sex erotic attraction with evolutionary theory is relatively easy. Evolutionists have a much more difficult time explaining the existence and persistence of same-sex erotic attraction. Symons (1979/1987) devotes only a tiny portion of his classic text, *The Evolution of Human Sexuality*, to same-sex erotic behavior. He assumes that same-sex eroticism is maladaptive, and yet hypothesizes that "gay" men and "lesbians" are *uncompromised* males and females, respectively—uncompromised, that is, by other-sex mate choice and intrasex competition for other-sex mates. In other words, Symons believes that "gay" men and "lesbians" are *pure* versions of their sex, uncompromised by the demands and desires of the other-sex. However, he implies that same-sex erotic attraction is caused by early exposure to testosterone and is, therefore, a biological condition; he flatly denies the claim that culture or society cause same-sex erotic behavior. In the end, Symons simply disregards same-sex erotic attraction as maladaptive and uninteresting, without commenting on its persistence, complexity, or potential value. In contrast to the tremendous detail and scholarship of the bulk of his text, the coverage of same-sex erotic attraction is pitiful.

Good evolutionary theories of general human sexual attraction must account for the persistence of exclusive and nonexclusive same-sex erotic attraction in ways that Symons has not. In the following sections, four major evolutionary-based theories of same-sex erotic attraction are reviewed: *sexual frustration, reproductive altruism-kin selection, personality* theories, and *heterozygous advantage*. These theories view sexuality as a function of selective mating strategies and describe sexual attraction as a conflux of various biological and social forces. More than the theories already discussed, evolution-based theories of sexuality emphasize the interaction between biology and the social environment. Unfortunately, many of the theories presented here are narrowly focused, illogical, fanciful, or incomplete. Still, it is worthwhile to discuss weak evolutionist theories in order to understand what is needed for a good practical general theory. A more comprehensive evolutionist account of the origin, function, and development of varied sexual attractions is proposed in chapter 5.

4.2.4.1 Overview of Evolutionary Theory and Sex Differences

Because there is much confusion about evolutionary theory in general, a brief overview may be helpful. Charles Darwin's (1859/1958) theory of natural selection holds that animals within a species vary in many ways. (As noted in the beginning of this chapter, diversity and variability are the norm in nature.) Darwin stated that over time some of these variations are consistently passed down to offspring in each generation. Animals that reproduce sexually pass on their inherited qualities to their offspring, and variations that are inherited are part of the evolutionary process. If on the average animals with particular characteristics produce more surviving offspring than do animals without those characteristics, animals with the traits are said to have a *differential reproductive advantage*. Darwin theorized that mechanisms for survival—for example, sweat glands to regulate body temperature—and reproductive competition are two evolved characteristics of animals, including humans. Long juvenile developmental periods, especially characteristic of modern humans, may be another naturally selected trait.

Natural selection contributes to sex differences. More commonly, however, Darwin (1871/1981) believed that reproductive competition within the species encouraged differential sexual selection of particular traits, and ultimately, sex differences. *Sexual selection* is the favoring of particular traits that improve reproductive success through competition for mates and mate choice. In other words, natural selection favors those biological traits that enhance survival and reproduction against predators, parasites, disease, environmental conditions, and threat from one's own species. Sexual selection favors those biological traits that influence differential reproductive success in the face of intersex choice of mating partners and intrasex competition for access to mating partners (Geary, 1998). These are the major forces behind evolution.

Among sexually reproducing species, genetic variability—individual differences—are desirable because of their potential contribution to survival against changing ecologies and fierce intersex and intrasex competition (Geary, 1998). Most sexually selected sex differences are the product of female choice of mating partners and male-male competition. Sexually selected sex differences in humans include body size and shape, secondary sex characteristics, physical strength, physical aggression and social dominance, sexual arousal patterns, coalition versus community building, parental investment, sexual versus emotional jealousy, parental care, and cognitive abilities. Traits that enhance reproductive success are advantageous. For example, investment in parental care increases offspring survival and directly enhances one's own reproductive success (Buss, Haselton, Shackelford, Bleske, & Wakefield, 1998). Traits that indirectly improve genetic survival are also advantageous. Contemporary evolutionists assert that *inclusive fitness* is enhanced if an organism's inherited qualities are passed on even without the organism directly producing offspring (Hamilton, 1964; Symons, 1979/1987). Thus, an individual who shares resources with brothers, sisters, or other close relatives contributes to the reproductive success of kin with whom he shares many of the same genes. Inherited traits that directly or indirectly facilitate reproductive success by solving an adaptive problem—for example, physical size or visible displays of virility— are called *adaptations* (Buss *et al.*, 1998; Dawkins, 1996). Other-sex erotic attraction or undifferentiated eroticism may be an adaptation if they enhance mate selection, sexual contact, reproduction, or survival. Unless same-sex erotic attraction has similar benefits, it is not an adaptation.

Still, traits can reliably persist with no benefit to reproductive success, although they have other benefits (Symons, 1979/1987). All behaviors are not reproductive or even adaptive. Same-sex erotic attraction, for example, could be a *by-product* of an adaptation or *residual noise* (Buss *et al.*, 1998; Dawkins, 1996). By-products are characteristics that do not solve adaptive problems and do not have functional design. By-products are coupled with and carried along by characteristics that are functional. In this sense, same-sex erotic attraction could be a by-product of other-sex attraction that directly affects reproductive success. Or same-sex erotic attraction could be conceptualized as noise or random effects produced by genetic variability and irrelevant to reproductive success. However, because exclusive same-sex erotic attraction both occurs with a high degree of regularity and yet reduces the likelihood of reproductive sex, it is probably not simply noise.

Same-sex erotic attraction could be an *exaptation* or a *spandrel* (Buss *et al.*, 1998; Gould, 1991, 1997). An exaptation is a feature that evolved for one function but has been co-opted for another function. Stephan Jay Gould (1991, 1997) views the human brain as a prime example of an exaptation. Other examples of exaptations are religion, language, writing, reading, principles of commerce, and the fine arts. An exaptation is currently functional. Same-sex eroticism could be an exaptation if it promoted social relationships, aided in the accumulation of resources, and enhanced survival at any time in the past. On the other hand, a spandrel is a functionless trait that arose as consequence of another functional feature. Spandrels are likely to be the by-products of a large brain. Examples of spandrels include awareness of one's mortality (Gould, 1991, 1997), wine making, doodling, and love of dogs. In this sense, same-sex erotic attraction could be an unexpected happenstance of a large mammalian brain that permits cognition, self-awareness, a wide range of feelings, and engagement in volitional nonreproductive sexual behavior. When a spandrel is co-opted for an adaptive function, it becomes an exaptation. Therefore, it is possible that same-sex erotic attraction developed as a functionless spandrel but later proved to enhance survival or reproductive success and became an exaptation. Evolutionist theories that explain same-sex eroticism as functional identify it as an adaptation or an exaptation.

Clearly, other-sex and same-sex erotic attractions have persisted at least through recorded human history. Evidence of individual preferences in mate selection and exclusive same-sex erotic attraction among non-human primates and other species suggests that varied sexual attractions among humans are not a recent phenomenon for our species. Of course, the forces that resulted in the selection of sexual traits occurred a very long time ago and may no longer be viable (Allman, 1994). Traits that developed as adaptations may not be functional today. Behavior in present society and current threats to survival cannot be used to explain the development or function of exclusive or nonexclusive sexual attractions. If these are selected traits, exclusive and nonexclusive sexual attractions are effects of ancient problems. Therefore, explanations for the existence of varied sexual attractions must be formed in terms of mechanisms that enhanced *past* fitness (Buss *et al.*, 1998). Modern life is very unlike the way in which our hominid ancestors lived, and modern life is a small blip in the history of hominid existence. Much of early hominid life, where sexually selected traits evolved, took place on large savannas in small sex-segregated collectives of kin and nonkin, where males spent their time hunting and fighting and females spent their time foraging for food and raising children. In this social context, reproductive competition between and among the sexes led to the selection of inherited sex-typed traits. Sex differences are, thus, the effects of ancient differential mating strategies, of which varied sexual attractions may be one result. A brief review of selected sex differences will facilitate the discussion of exclusive and nonexclusive sexual attractions as potentially evolved traits.

Psychologist David Geary (1998) provides an excellent overview of the evolution of human sex differences. He notes that female-female and male-male competition, as well as mate choice, fueled the development of sex differences. Social and ecological circumstances also influence the nature of these forces. For example, Geary explains, a low number of adult males in a community, perhaps as a result of death in warfare, increases female-female competition for sexual partners, while the reverse is true if the community has a small female-to-male ratio. Socially imposed conditions, such as monogamy or polygyny, also pressure intrasex competition and intersex mate choice. In societies with socially imposed monogamy, females have more freedom in mate choice and male-male competition is less intense. From an extensive review of the literature, Geary lists several sex differences that are reliably evident across cultures. On the whole, boys and men are more physically aggressive than are girls and women. When aggressive, girls and women are more likely to use indirect methods such as gossiping or shunning rather than physical force. Boys are more physically active than are girls (Bem, 1996). Both males and females seem to readily form in-groups and out-groups, but the sexes maintain them in different ways and for different reasons. In general, boys and girls, as well as men and women, prefer activities with others of the same-sex—one kind of in-group (Geary, 1998). From an early age across a variety of cultures, children spend most of their time in same-sex play, and this pattern continues and even increases in adulthood. However, boys tend to play in large groups and prefer games with clear winners and losers, whereas girls prefer to play in small groups without defined hierarchies and like games without clear winners. Boys and men prefer instrumental same-sex coalitions organized by social or political hierarchies and maintained by threat of force, either directly through physical aggression or indirectly through verbal threats and name-calling. On the other hand, girls and women tend to form small, stable same-sex coalitions or dyads that are egalitarian and emotionally expressive. Dominance positioning is evident within female groups, but it is less direct and less often physically aggressive than in male groups. To summarize, Geary states that male-female sex differences are associated with traits that facilitated survival *within same-sex groups*, which most typifies the social environment of early hominids, and with traits that influenced effective mate choice, such as displays of health, fertility, dominance, and resources, as well as parental commitment. Such sex differences are not necessarily biological traits; they are, however, traits possessed by our ancestors and passed on to us, either socially or biologically.

Michael Bailey and colleagues summarized a wealth of data about evolutionary-based sex differences and their relationship to sexual attraction. Bailey and colleagues noted first that males in the past maximized their mating opportunities by being interested in casual, uncommitted sexual encounters. Consistent with this supposition, men report more frequent desire for sex with more novel partners than do women (McConaghy, 1993). Men also are more willing than women to engage in casual and impersonal sex (Bailey et al., 1994). According to this line of thought, because women commit greater resources toward reproduction, they gain little benefit from frequent casual sexual activity, unless an occasional extra-pair copulation with a genetically superior male enhances reproductive success (Geary, 1998). Bailey points out that like "heterosexual" men "gay" men also desire a high frequency of sexual activity with many sexual partners. However, "lesbians," report a low desire and very low frequency of sexual activity. Bailey and colleagues (1994) speculate that the higher frequency of sexual activity among "heterosexual" women compared with "lesbians" may be related to accommodating demanding male partners and does not represent "heterosexual" women's true desired frequency of intercourse. An alternate explanation is that because women in general are socialized to be less assertive than men in initiating sexual activity, the frequency of sexual activity among "lesbians" reflects a social constraint on their frequency of sexual activity.

Second, Bailey and colleagues (1994) noted that males benefit strategically by being readily aroused sexually, whereas females do not experience similar benefits. Consistent with this prediction is evidence that men are easily aroused sexually to visual stimuli and often sexualize social encounters. The unquenchable depths of male sexual appetite and responsiveness are no better illustrated than by the multibillion dollar "heterosexual" male pornography industry. In spite of several attempts, only a small weak market for female "heterosexual" pornography has ever existed. Although women have less disposable income than men to spend on pornography, this inequity does not explain the dramatic differences in sexual response. Consistent with these sex differences, "gay" men and "lesbians" show similar behavioral patterns. The large male-male pornography industry has completely overshadowed the few erotic materials that are directed toward "lesbians." Feminists loudly decry the small but vocal prosexual "lesbians" who advocate female-female pornography, sexual freedom, and consensual sadomasochistic sexual play. Furthermore, males and females differ in the content of erotica. In general, male-oriented erotica illustrates nude bodies, sexual organs, and impersonal sex acts, whereas female-oriented erotic materials more often describe sexual acts in romantic relationships and emphasize the partner's non-sexual characteristics (Zilbergeld, 1992).

Third, Bailey *et al.* (1994) noted that evolutionary theory predicts that because females have a larger and more focused reproductive investment, they are more concerned than males with a partner's social status and his ability to provide protection and acquire resources that promote the survival of offspring. On the average, females are attracted to older, well-established, successful males. Males, however, are more interested on the whole with securing a willing attractive sexual partner than a socially prominent one. When "heterosexual" males are selective, they prefer younger female partners who possess characteristics associated with fertility such as youth, symmetrical body shape, and ample hip-to-waist ratio (Bailey *et al.*, 1994; Gangestad & Thornhill, 1997; Thornhill & Gangestad, 1994). By the same token, because males are fertile for most of their adult lives and because some time is required for males to achieve social position, youth is not a major criterion in a mate for females. Other characteristics of health such as body symmetry, vigor, and prominent secondary sex characteristics are important. Consistent with these suppositions, both "heterosexual" and "gay" men rate younger individuals as most sexually attractive, and "heterosexual" women rank older men as more attractive than younger men. "Lesbians," however, do not seem to connect partner's age to attractiveness. Bailey and colleagues point out that according to evolutionary theory men more than women should value and respond to a mate's physical attractiveness. This relationship is also supported, and "gay" men and "lesbians" demonstrate the expected sex differences.

Fourth, and finally, Bailey *et al.* (1994) reported that because of intense intrasex competition for status and partner, as well as uncertainty about paternity, males should be more distressed than females by a partner's sexual infidelity. Likewise, because of greater commitment to parental care and the resources needed for offspring survival, females should be more concerned about the emotional fidelity of her mate. Indeed, men report more sexual jealousy than do women (Geary, 1998). Illustrative of this fact, institutionalized mate-guarding is a common feature of male-female relationships across cultures. Women, on the other hand, are more likely to react to their partner's unstable emotional commitment than to his sexual infidelity. This particular sex difference is less evident among "gay" men and "lesbians" than among "heterosexuals." "Gay" male couples, for example, sometimes have extra-pair or shared sexual partners with a low risk of sexual jealousy (McWhirter & Mattison, 1984). For male couples who agree to have other sexual partners, their emotional commitment is to the primary relationship. On the other hand, among "lesbians," extra-pair sexual encounters are less common but not unknown (Blumstein & Schwartz, 1983).

With this brief overview of evolutionary theory and summary of sex differences as a result of evolutionary forces, we are in a better position to evaluate evolutionary theories of same-sex erotic attraction. Although many of these models are incomplete or logically inconsistent, chapter 5 applies a more comprehensive evolutionary theory to explain the origin and development of varied sexual attractions.

4.2.4.1 Sexual Frustration

A conceptually weaker model claims that "gay" men are inadequate, unskilled, or damaged "heterosexuals." Gallup and Suarez (1983), for example, describe "gay" men as frustrated "heterosexuals." Within this model, male-female mating strategies are seen as a series of strategies and counter-strategies designed to improve one's own reproductive success (McKnight, 1997). Gallup and Suarez (1983) purport that in general females delay sex in order to make the best choice of father for their children and because their investment in raising children to self-sufficiency and reproductive age is large. Female sex-delaying strategies advantaged males who were demanding, overly sexual, and seductive. According to Gallup and Suarez, the female strategy to protect their reproductive investment resulted in a male counterstrategy to be more sexual and persistent. Since young men are not particularly patient or skillful or able to access many receptive females (who desire older, more successful men anyway), sexually frustrated young men turn to each other. Male-male sexual behavior is then reinforced through practice because it is readily available and does not require the emotional entanglements or skillful negotiation that is demanded for sexual activity with females. In other words, Gallup and Suarez assert that "gay" men are sexually frustrated and inept "heterosexuals." However, they point out that sexual frustration and clumsiness cannot be the cause of same-sex female erotic behavior since mate choice is determined more often by females. This model offers no theory about female-female attraction.

Although Gallup and Suarez's (1983) sexual frustration theory may explain the frequency of same-sex erotic activity among adolescent males in many cultures, there is no evidence from the literature that "gay" men are socially inept, unable to obtain other-sex partners, or sexually frustrated. In fact, on the average they report more lifetime sexual partners than do "heterosexual" men (Laumann et al., 1994), which suggests that "gay" men have many social skills. Furthermore, at least in the past a significant proportion of "gay" men married women (Bell & Weinberg, 1978) and, currently, it is estimated that 40 to 60% of "gay" men are in male relationships (Kurdek, 1995). Sex therapists Masters and Johnson (1979) even claimed that "gay" men and "lesbians" are more skillful than "heterosexuals" at love-making. All said, little evidence supports the notion that "gay" men lack social skills or are poor-quality "heterosexuals."

4.2.4.2 Reproductive Altruism and Kin Selection

A third major evolutionary account holds that a limited number of same sex-oriented individuals provide a reproductive advantage to kin, who somehow manipulate the direction of sexual attraction in relatives. Sociobiologists Edward O. Wilson (1975, 1978), Robert Trivers (1971, 1974, 1985), and James Weinrich (1987a,b; Kirsch & Weinrich, 1991) maintain that the persistence of same-sex erotic attraction in the population suggests that the trait is biological and that it confers some survival or reproductive advantage either to the individual or to their kin. A hallmark of sociobiology is the assumption that traits persist because they are adaptive. E.O. Wilson (1975, 1978) hypothesized that parents pass on genes associated with same-sex erotic attraction to offspring, because doing so benefits their other reproducing offspring. Although at first glance parents appear to reduce their reproductive chances by limiting their number of reproducing offspring, Wilson argues that parents' genes benefit because sibling competition for mates and resources decreases. Wilson claims that nonreproducing individuals direct their time and effort to raising their siblings' children (who share some of their genes) to reproductive age, ensuring their own genetic survival. Rather than further dilute scarce environmental resources, a few altruistic nonreproducing offspring who help their siblings would serve parental interests by maximizing the reproductive success of some offspring. Although the idea of reproductive altruism is intriguing, it suggests that individuals with same-sex erotic interests are more altruistic than are people with other-sex interests and that same-sex oriented persons assist their kin more often than do "heterosexual" kin (Kirsch & Weinrich, 1991). However, current published research data provide no support these suppositions. In fact, several studies suggest that "gay" men prefer to invest their resources with unrelated "gay" men rather than with kin (Hewitt, 1995).

Trivers and Weinrich have also argued alternately that in societies in which marriage is mandatory parents may benefit by having offspring with same-sex erotic interests. In many cultures, parents pick or have great influence in the selection of a marriage partner for their child. Trivers and Weinrich suppose that individuals with some same-sex erotic interests would be less resistant to marrying a partner picked by their parents. Furthermore, once married, individuals with same-sex erotic interests would be less likely to produce illegitimate children from extramarital sexual encounters. Thus, parents may favor children with same-sex erotic attraction (Trivers, 1971, 1974, 1985). Again, no published data support this hypothesis. And even if true, it is uncertain that married individuals with same-sex interests would reproduce more often and more successfully than other-sex oriented siblings, who might also dislike their marriage partner but produce numerous illegitimate children. From an evolutionist point of view, reproductive success means gene survival to successive generations, not the endurance of family names or fortunes. What is more, sociobiologists who advocate the reproductive altruism or kin selection models fail to explain how parents influence the direction of their children's sexual attraction.

4.2.4.3 Personality

Personality theorists have employed evolutionary theory to assert that forms of sexual attraction are by-products of nonsexual traits. Three personality theories of sexual attraction are discussed here.

William Byne and Bruce Parsons (1993) hypothesize that biologically based personality traits related to sex differences facilitate the development of male-male sexual attraction. In short, sex-typed traits lead to the development of other-sex erotic attraction, and cross-sexed traits lead to same-sex male erotic attraction. (As an aside, this theory places the cause of sexual attraction in the cortex rather than the hypothalamus). Byne and Parsons claim that a high frequency of cross-sex behavior or *gender nonconformity* among young boys who later identify as "gay" supports their explanation. They assert that *male-typical* traits are novelty seeking, low harm avoidance, and reward-independent behavior; *female-typical* traits are low novelty seeking, harm avoidance, and reward dependence. Thus, according to Byne and Parsons, boys who inherit low novelty seeking, high harm avoidance, and reward dependence are unlike typical boys, feel different from their male peers, and are more open to nonsexual female-typical experiences that in turn facilitates the development of same-sex erotic attraction. Essentially, Byne and Parsons say that typical girl-behavior and typical boy-behavior lead to other-sex erotic attraction in adulthood. If boys have social experiences that are more similar to girls, then those boys are likely to grow up liking males. Curiously, Byne and Parsons suggest that their personality theory of sexual attraction applies equally well to the development of same-sex eroticism in females, although they do not describe this process. Even so, the idea that "gay" men—who have more lifetime sexual partners on the average than "heterosexual" men—are low in novelty seeking and high in harm avoidance is laughable.

Nathaniel McConaghy (1993) starts with a premise similar to that of Byne and Parsons. He assumes too that typical sex-typed traits lead to other-sex erotic attraction and attributes cross-gendered personality traits—"sissiness" for males and "tomboyism" for females—to prenatal exposure to cross-sex hormones. Thus, when males are exposed to too much estrogen and females are exposed to too much androgen, they behave in ways like the other sex. According to McConaghy, parents and peers respond to cross-gender behaving individuals with negative reactions and disapproval but, rather than redirect learning towards rewarded behavior, negative reactions to cross-gendered traits promote atypical learning. Persistent atypical behavior and an atypical learning milieu contribute to the interpretation of erotic feelings as same-sex attraction, in McConaghy's view. This is an important point because McConaghy believes that cross-gender behavior *precedes* same-sex erotic feelings. He is so wedded to this idea that when "gay" men in his studies failed to report cross-gendered histories and negative social reactions he assumed that they were lying or simply giving socially desirable responses. Somewhat surprisingly then, he notes that only a few people rigidly conform to sex-typical roles and that most people have a combination of sex-typed and cross-gendered erotic traits. To his credit, McConaghy hypothesizes that degrees of same- and other-sex erotic attraction probably provide flexibility in complex social relationships and in society.

Lastly, Daryl Bem (1996) also starts with the premise that sex-typed personality traits promote the sexualization of early social experiences, which later are interpreted as other-sex erotic attraction. He notes that childhood social experiences are mostly with same-sex and same-age peers and suggests that through play with same-sex peers children develop an autonomic familiarity toward and comfort with individuals of the same sex. However, other-sex persons, being less familiar, evoke fear and anxiety in children and increased arousal. During puberty, autonomic arousal to contact with other-sex individuals is translated into and perceived as sexual arousal or erotic attraction. Bem calls this process the *exotic-to-erotic* theory of attraction. Unlike most theorists, he also describes the development of other-sex erotic attraction. According to the exotic-to-erotic theory, children with cross-gendered personality traits associate more often with other-sex children and spend less time with same-sex playmates. Through extended play, autonomic arousal of children with cross-gendered traits habituates to other-sex peers. On the other hand, infrequent contact with same-sex peers evokes fear and anxiety. During puberty, autonomic arousal around same-sex persons is perceived as sexual arousal and attraction. Same-sex individuals are sexualized and experienced as sexual objects. In short, Bem hypothesizes that boys with cross-gendered traits sexualize the masculine traits of "heterosexual" boys and men—the very traits that "gay" men supposedly lack. According to this model, "gay" men identify with the masculinity of "heterosexual" men but masochistically settle for relationships with other "gay" (effeminate) men. Said another way, Bem's model asserts that "gay" men have romantic relationships with other "gay" men, although they really yearn for relationships with "heterosexual" men. More troubling, however, if "gay" men eroticize "heterosexual" men, why would they find "gay" men, who are supposedly less masculine, sexually attractive at all? A similar conclusion can be drawn for "lesbians" who, according to the exotic-to-erotic theory, should be attracted to "heterosexual" women not to other "lesbians;" "lesbians" should be too masculine to evoke sexual arousal in each other.

Although creative and interesting, these personality theories of sexual attraction present a number of difficulties. First, there is no published evidence that people with same-sex erotic interests have more biologically based cross-gendered traits than other-sex oriented individuals. It is true that on the average "gay" men and "lesbians" report more sex-atypical behaviors in childhood than do "heterosexuals" (Bailey, Nothnagel, & Wolfe, 1995; McConaghy, 1993). But is the frequency of cross-gendered behaviors more important than the type? Sex-atypical behaviors could also be social, rather than biological, in origin. If we assume for the moment that these personality theorists are correct, "gay" men *should* act like females and not just include a few female-typed behaviors in their repertoire. To test this supposition, Gladue and Bailey (1995) compared 182 "heterosexual" men, 68 "heterosexual" women, 74 "gay" men, and 77 "lesbians" on sex-typical characteristics, matching for age, ethnicity, and education. Not surprisingly, sex differences were clear. On the whole, men more than women described themselves as aggressive and interpersonally competitive. However, compared to "heterosexual" men, "gay" men were more variable on cross-gendered traits than were "lesbians" relative to "heterosexual" women. Although "gay" men reported being less physical aggressive than did "heterosexual" men and were more comparable in aggression levels with "heterosexual" women, "gay" and "heterosexual" men did not differ on verbal aggression or competitiveness. In addition, "heterosexual" women and "lesbians" evidenced no differences on competitiveness, physical aggression, or verbal aggression. Gladue and Bailey concluded that while sex hormones may promote sex differences in physical aggression, social factors are more likely the cause of differences in verbal aggression and interpersonal competitiveness; thus, cross-gendered or sex-atypical characteristics are weakly related at best to sexual attraction. It is well know, for example, that changes in social status influence levels of testosterone in men, which may alter social behavior (Booth & Dabbs, 1993; Booth *et al.*, 1989). Frequent testosterone-suppressing experiences such as ridicule and social rejection might facilitate the development of less sex-typical behaviors, independent of sexual attraction.

The second major criticism of the personality theories discussed here is that they simply interpret gender role behavior as sexual orientation. In this view, "heterosexual" men are masculine and do male-typical things, and "heterosexual" women are feminine and do female-typical things; "gay" men do not do male things—that is, love females—and, therefore, are less manly or effeminate. By the same logic, "lesbians" are masculine and not-feminine. The problem, of course, is that the relationship between sex-atypical behaviors and same-sex erotic attraction is weak. However, the biggest problem with cross-gendered trait theories of attraction is that they fail to explain how individuals with sex-typical traits develop same-sex erotic attraction and why people with cross-gendered traits do not (Bailey *et al.*, 1995). Bem (1996) is the exception. Indeed, many "gay" men have masculine traits, and many "lesbians" are feminine. Likewise, some "heterosexual" men are effeminate, and some "heterosexual" women are masculine. Which traits then are important and when? The personality trait theories give equivalent significance to all cross-gendered traits.

Lastly, these personality theories are essentially reworked androgen under- and overexposure theories of sexual orientation that suffer the same limitations that were discussed earlier. These theories closely link sex-typed traits to prenatal sex hormones, specifically, testosterone. However, it is likely that nonsex-typed traits have a greater influence on development of sexual attraction than sex-typed ones

4.2.4.1 Heterozygous Advantage

Another interesting explanation for same-sex erotic attraction is that it persists because it confers a selective advantage. Given that sexual attraction is a product of multiple genes—and this is one of the least controversial suppositions by biological theorists—population geneticists argue that selection should favor a *balanced* or *stabilizing optimal type*, which is an intermediate form of the given trait (McKnight, 1997; Ridley, 1994). According to this argument, if other-sex erotic attraction is a dominant version of the trait and same-sex attraction is a recessive version, then neither *homozygous* versions of the trait—two dominant alleles or two recessive alleles—provide as much genotypic diversity or potential reproductive advantage as *heterozygous* versions—one dominant and one recessive allele.

Let me give background to this argument by describing theoretical biological models of sexual attraction. Again, *genes* are functional segments of *deoxyribonucleic acid* (DNA) molecules that directly or indirectly influence the expression of a trait. Anne Fausto-Sterling (1992) provides a good functional definition of genes as "molecular templates consisting of nucleic acids that replicate with the aid of enzymes. These templates provide the code … for the synthesis of specific proteins" (p. 65). DNA then is the "cookbook" for an organism's development and form. In this discussion, genes associated with sexual attraction are described in the abstract sense. No genes "for" sexual orientation have been identified.

The simplest biological model of sexual attraction requires no genes associated with eroticism. In this model, sexual attraction is a by-product of typical exposure to fetal sex hormones during sex differentiation. Sexual attraction and sex are inextricably linked—androgens initiate male development and produce attraction to females, and the absence of androgen permits female development and produces attraction to males. Because female development is the default physiological form, sexual attraction to males should be the default erotic orientation. We have discussed similar ideas earlier in this chapter. According to the no-gene attraction model, a man who likes men is not fully male; he has female characteristics and is cross-sexed for sexual attraction. Similarly, a woman who loves women is not a real woman. Sexologist Karl Heinrich Ulrichs in the 19th century purported that men and women who loved members of the same-sex were psychologically cross-sexed (Greenberg, 1988).   John Money's (1988) neurohormonal theory of gender cross-coding is also consistent with this idea. Money's theory effectively translates Ulrichs' 19th-century model into contemporary medical verbiage.   Again, the problem with cross-sexed hormone models is that there is no strong evidence to support it.   Physical differences between "gay" and "heterosexual" men and between "lesbians" and "heterosexual" women are negligible. "Gay" men and "lesbians" are not more prone to hormonal anomalies than are "heterosexuals." And there is more intrasex variance in circulating hormones than there is difference between sexual orientations in circulating hormones.   What is more, this model cannot account for people with erotic interest in both sexes. So-called bisexuals would be hermaphrodites—both male *and* female—because in this model sex is linked to attraction. In short, the no-gene model is not viable.

The next simplest biological model of sexual attraction is similar to the first. In this case, prenatal sex hormones either switch "off" or leave "on" a single gene for sexual attraction. Sex-typical hormone exposure throws the switch off or leaves it set on the default attraction to males. This model simply repeats the same conceptual problems as the previous one.

A third model hypothesizes that two forms of the trait exist—a dominant version (other-sex erotic attraction) and a recessive one (same-sex attraction). Everyone gets two copies of the gene. Two dominant alleles produce exclusive other-sex erotic attraction, and two recessive alleles produce exclusive same-sex attraction. Mixed alleles result in the dominant form of sexual attraction being expressed. Although individuals who have same-sex erotic attraction would not reproduce, other sex-oriented individuals with the recessive trait could pass it on to offspring. This model views sexual attraction as a dichotomous trait that is either on or off rather than as a continuous trait. This model also fails to account for people who have erotic interests in both sexes, which at least for males has been the historical norm.

However, a variation of this model nicely accounts for variability in sexual attraction. (This heuristic example in no way should imply that a single gene determines sexual attraction.)   Again, everyone receives two copies of the gene—a dominant version and a recessive one. However, in this model, having mixed genotypes is optimal. In this model, having a gene associated with a trait does not indicate if or to what degree the trait will be expressed. Thus, in this model, sexual attraction could be a continuous trait. Consequently, other biological factors as well as social experiences influence the strength and form of the trait. This model predicts variety in sexual attractions across individuals. If mixed genotypes or heterozygous genes are optimal, then exclusive forms of attraction—other-sex and same-sex—are less advantageous to ultimate reproductive success.

Now back to the heterozygous advantage model mentioned at the beginning of this section. Having both versions of a trait is an advantage over having two copies of the same version of the trait, because nature favors genetic and behavioral variability. Genes associated with other-sex erotic attraction also direct or are related to abilities that are different from abilities associated with same-sex attraction. If one function of the dominant allele is to promote other-sex erotic attraction, possible functions of the recessive allele could be to enhance general sexual desire, or physical attractiveness, or social skills (McKnight, 1997). A by-product of having a recessive allele is same-sex eroticism.   Thus, individuals with both versions of the trait (heterozygous genotypes) would be more adaptive and socially versatile than individuals with just the dominant or just the recessive version. Said another way, having the same-sex eroticism recessive gene would somehow make sexual encounters with the other-sex more probable or perhaps improve social relationships or resource-gathering skills that enhance desirability as a mate or facilitate the survival of offspring to reproductive age. Heterozygous alleles for sexual attraction may also produce a heightened sex drive.   The persistence of heterozygous genotypes of eroticism is a function of female mate choice; that is, males with both versions of the sexual attraction trait should be perceived as more attractive or more successful than males with only exclusive erotic interests.

Are heterozygous genes really optimal? In a changing environment, polymorphic genotypes have a distinct advantage because they enhance adaptability (McKnight, 1997).[6] Heterozygosity contradicts the idea that there can be one and only one optimal genotype in a species (Ruffie, 1986). *Directional selection* favors a single optimal type that eliminates "bad" recessive genes from the population over time (McKnight, 1997).   In evolutionary terms, genes are bad if they reduce reproductive fitness.   In stable environments, homozygosity is an advantage (Wallace, 1968).   Even so, bad genes are never completely removed from the population, and the optimal homozygous type is elusive.   Furthermore, environments are rarely stable.   Population geneticists note that stabilizing or balanced selection favors an intermediate heterozygous type, a strategy that is more consistent with human evolution (McKnight, 1997). *Balanced selection* assumes that having one copy of the recessive gene confers some reproductive advantage to its host. According to this theory, having two copies of the recessive gene results in exclusive same-sex erotic attraction and no reproductive advantage. In addition, individuals who have two copies of the dominant gene—other-sex erotic attraction—are less reproductively fit than people with discordant copies.   Thus, given a changing, challenging environment, the presence of recessive genes for same-sex attraction might confer a reproductive advantage and maintain genotypic diversity (McKnight, 1997).   Balanced selection of heterozygous traits also works if there is more than one optimal type operating in the population (Ruffie, 1986).   McKnight emphasizes that it is the *associated characteristics* of a recessive gene for same-sex erotic attraction are reproductively advantageous and not same-sex attraction itself. Such characteristics could include a high sex drive, signs of virility, social and verbal skills, and self-confidence, all of which could differentially influence female mate choice.   If indeed multiple genes affect sexual attraction, then individuals could receive large or small doses of particular types of sexual attraction. Dose-dependent genetic loads would create a range of varied sexual attractions and associated characteristics.

Although McKnight (1997) views sexual attraction as a sex-linked trait residing on the sex chromosomes, he believes that this *balanced heterozygotic fitness model* applies only to males and that same-sex attraction in females has a different etiology. Hamer and Copeland (1994) describe one possible heterozygotic fitness model of sexual attraction for women, which they call the *overloving mother* theory. They hypothesize that a gene on the X chromosome related to same-sex attraction may boost female sex drive. Consequently, females with a high dose of this trait could produce more children than other women. Such women should also have more sons with a same-sex erotic orientation than would women without the same-sex erotic trait. Thus, the reproductive advantage is for the mother, not her sons. In Hamer and Copeland's model, same-sex male attraction is an accidental by-product and not a selected trait. For males, any allele on the X chromosome is dominant because there is no corresponding allele to mask it on the other chromosome (Hodson, 1992). Most X-linked traits are recessive in women, who act as carriers. However, the trait becomes dominant if two X-linked versions are received from both parents.

In short, heterozygous genotypes for sexual attraction, at least for males and possibly for females, represent a reproductive advantage in unstable, challenging environments. A combination of multiple genes associated with sexual attraction and heterozygous genotypes permit a range of sexual attractions and associated characteristics. In this model, characteristics associated with mixed genotypes, including erotic attraction to both sexes, confer a selective advantage to their hosts. It is unlikely that the process that produces same-sex attraction in men is the same for women. The number of women who report same-sex erotic attraction is about half that of men. The sexes may differ in which genes or what combination of genes contribute to sexual attraction.

Thus, the heterozygous advantage model explains exclusive same-sex attraction as a by-product of genetic variability and proposes that males and possibly females who have both versions of the trait have a selective advantage over individuals with single versions of a trait (McKnight, 1997). One difficulty in testing this hypothesis is the reluctance of men in Western culture to acknowledge *any* same-sex erotic interest. Still, the theory has rational merit and avoids the conceptual problems that plague other evolutionary-based models.

4.2.5 Summary of Findings from Biological Research

A pattern of circumstantial data suggests that at least male sexual attraction and probably female sexual attraction have a biological foundation. Findings from twin and family studies of "gay" men are consistent with same-sex erotic attraction as an inherited trait. Across people of different sexual orientations, neuroanatomic and cognitive functional differences also support this possibility. Still, sex differences are much more evident than are sexual orientation differences. Positive findings from two of three studies comparing genetic markers among "gay" brothers also support the contention that one form of same-sex male attraction is genetic. However, assumptions of a linear androgen underexposure theory fail to account for variability in sexual attraction among men.

For women, evidence of a biological component of sexual attraction is more tenuous. Twin and family prevalence studies of same-sex erotic attraction among women are less strong than for "gay" men. Neuroanatomic and cognitive differences between "heterosexual" and "lesbian" women are not predictable, with the singular exception that "lesbians" demonstrate otoacoustic emissions from the inner ear that are more similar to men than "heterosexual" women. Sex differences in otoacoustic emissions result from prenatal exposure to androgens. Thus, androgen may play a small but specific role in female sexual orientation. However, the conventional androgen overexposure theory fails to account for varied sexual attractions among women.

A major problem in biological research on sexual attraction can be attributed to conceptual errors and unexamined assumptions. Many biological theorists conflate *sex, gender, gender roles, sexual behavior*, and *sexual identity* (Kauth & Kalichman, 1995). When this happens, sex or gender roles are assumed to lead *naturally* to "heterosexual" erotic attraction, and nonheterosexual attraction is thought to be due to cross-gendered traits and behaviors, even though evidence does not support this view. Biological researchers also constrain the quality of their data when they value sexual behavior as more valid than sexual feelings and fantasies, even though sexual feelings and sexual behaviors do not always correspond (McConaghy, 1993). Biological researchers almost never assess sexual feelings, let alone compare sexual feelings with behavior. "Heterosexual" adults typically report more variety in sexual feelings than in sexual behavior (McConaghy, 1987; McConaghy, Buhrich, & Silove, 1994). Let me give one example of how a bias toward sexual behavior can misdirect researchers. Van Wyck and Geist (1984) attempted to provide a better estimate of "heterosexual" and "homosexual" orientation by reanalyzing some of Kinsey's data. They computed separate sums of "heterosexual" and "homosexual" behavior that were divided by total sexual experience for 3,849 men and 3,392 women. Based on sexual behavior, they concluded that people are predominantly "heterosexual" and that sexual orientation is extremely bipolar, with almost no one reporting sexual activity with both sexes. Although Van Wyck and Geist also computed scores for sexual feelings, they did not report these data, presumably because they were less bipolar or just less clear, which is much more interesting! Thus, an emphasis on sexual behavior to the exclusion of sexual feelings omits important information about sexual attraction and misrepresents sexual orientation.

Finally, the static androgen-threshold theory of sexual attraction is overly simplistic and fails to account for variability in timing, duration, and exposure to androgens, as well as competition with estrogens for receptor sites. For the most part, such theories describe androgen effects on sexual attraction as fixed rather than as predispositions. In summary, conventional biological theories of sexual attraction are in need of salvaging and refitting, or abandonment.

4.3 SALVAGING BIOLOGICAL MODELS OF SEXUAL ATTRACTION

As social constructionists have rightly noted, biomedical theories are based on assumptions, and too many biological researchers seem unaware of the flawed assumptions underlying their conceptual models and research questions. Indeed, Science and scientists are influenced by cultural values. Although constructionists conclude that research studies have severe limitations and irreparable flaws, there is no advantage in throwing out baby Science because the bath water is dirty. Demands for the absence of imperfection and social bias in research is unrealistic and a standard that no academic discipline can meet. Science is a process and an imperfect one. Scientific conclusions are statements of probability, not certainty. However, I believe that helping scientists become aware of how cultural values infect and influence the practice of Science will sharpen research questions and strengthen methodology. Science is a process of discovery and self-correcting refinement through prediction, falsifiable hypotheses, and standardized methodologies. Although many people and some scientists desire Truth and Justice, these ideals will not be found through scientific pursuits. People who want absolute Truth and Justice will be more comfortable with theological explanations than with scientific ones and that is where they should look for answers. In spite of an imperfect system and imperfect world, scientific research into the biology of human sexual attraction has much to offer.

Humans are biological organisms and, as such, neurochemical processes underlie the detection and interpretation of all experience. Over countless generations of hominids, natural and sexual selection has favored particular traits that enhanced the survival and reproductive success of our ancestors. Consequently, it is reasonable to suppose that sexual attraction is a biologically based trait that provides some advantage in at least some form to its host. To deny the biology of sexual attraction is inconsistent with how our bodies work.

Biology is not generally a dictator. Genes set a range of possible expression (genotype), and where that individual finds himself or herself in that given range—the actual form of the trait (phenotype)—is a function of numerous biological and environmental interactions. As well, the expression of traits associated with sexual attraction is likely to depend on the strength of the traits, as well as future neurochemical events, environmental events, and social experiences at critical developmental intervals. From the discussion above, it seems most probable that a combination of genes creates various hormonal events that produce a potential *range* of attraction. In other words, a genetic contribution to sexual attraction is likely to be a predisposition to make certain erotic associations, given particular social experiences and available stimuli. As the mechanism of action for genes, prenatal sex hormones may establish under what social and hormonal conditions a given trait will develop (Thornton, 1986).

Richard Whalen (1991) argues that sex hormones regulate erotic meaning and perception of stimuli by altering "stimulus saliency" and "detectability." Based on non-human data, Whalen asserts that exposure to prenatal hormones predispose animals to detect and perceive particular stimuli as sexual. A similar process may occur in humans. Hormone levels are known to fluctuate with changing social events (Booth & Dabbs, 1993; Booth *et al.*, 1989), which may in turn modify perception and learning. The stimuli-processing environment of the brain is affected continuously by circulating sex hormones and other neuromodulaters. As a result, at any given time and place pleasurable responses may be facilitated, whereas pain and sorrow are suppressed. In healthy rats, for instance, sex hormones modulate the functioning of neurotransmitters and neuropeptides and promote dendritic growth and new neural connections (Kawata, Yuri, & Morimoto, 1994). What is more, learning experiences increase synaptic strength and the likelihood of similar responses (Frey & Morris, 1997; Rogan, Staeubil, & LeDoux, 1997). In male rats, sexual activity leads to enlargement of the spinal nucleus of the bulbocavernosus (Breedlove, 1994). Through behavioral experience, well-used neural pathways become stronger, more efficient, and more highly interconnected.

If these inferences are correct, sexual attraction is not likely to be found in a single brain structure. Sexual attraction is probably "located" in several brain sites and more accurately conceptualized as a network, rather than a locus. In addition, because brain tissue grows at different rates, particular regions may be differentially sensitive to hormone activation. Sequential or overlapping critical periods for one or more sex hormones at different sites could produce wide variation in sexual attraction, given a supportive social environment (Collaer & Hines, 1995).

A more sophisticated interactionist model of sexual attraction is needed—a theory that recognizes people as biological organisms, both acting on and acted upon by a powerful changing social environment. This is now the subject to which we turn.

CHAPTER 5: A SYNTHESIS: THE DEVELOPMENT AND FUNCTION OF SEXUAL ATTRACTIONS

Even a superficial look at other societies and some groups in our own society should be enough to convince us that a very large number of human beings— probably a majority—are bisexual in their potential capacity for love. Whether they will become exclusively hetero or exclusively homo for all their lives and in all circumstances or whether they will be able to enter into sexual and love relationships with members of both sexes is, in fact, a consequence of the way they have been brought up, of the particular beliefs and prejudices of the society they live in and, to some extent, of their own life history. —
Margaret Mead

There is nothing so practical as a good theory. — Kurt Lewin, *Principles of Topological Psychology*

From the preceding chapters it should be evident that single-factor models of human sexual orientation do not adequately account for persistent variability in sexual attraction. Many single-factor models such as the androgen under- and overexposure theories are overly simplistic and conceptually flawed. Furthermore, most of these models address only same-sex male attraction. Although multi-dimensional models of general sexual attraction are clearly necessary, interactionist models that are in current use often just extend or combine single-factor models, retaining the same conceptual flaws. In addition, most interactionist models describe only proximate causes of sexual attraction and say nothing about ultimate causation. However, a good practical theory of human sexual attraction should account for the origin and persistence of varied sexual attractions, as well as explain the neurohormonal-social experiential interaction that

169

facilitates the form and expression of sexual attraction. This chapter attempts to present such a comprehensive theory.

The first section of this chapter focuses on evolutionary or ultimate causes of varied sexual attractions. Contemporary human beings living in this highly technological, rapidly mobile, compartmentalized, individualistic and self-conscious age are as much products of antiquated biological and social forces as were our human ancestors. Whether we accept it or not, age-old biological forces influence how we live our complex modern, socially constructed lives. Sexual attraction is anchored in a biology that has been shaped by ancient social pressures on survival and reproduction, although few of these forces exist in the same forms today. Biological mechanisms and social forces intertwine and interact in the individual. This first section describes how sexual attraction evolved from an instinctive sex drive and how the sexes exploited varied sexual attractions to manage intrasex and intersex challenges.

The second section of this chapter focuses on neurohormonal-social experiential or proximate causes of varied sexual attractions. This section describes the neurohormonal mechanisms that establish erotic predispositions in social information processing systems and how erotic associations develop into stable templates and eventually into a sexual orientation. The new interactionist model of sexual attraction is contrasted with a currently accepted model of attraction.

Now, to begin, let us go to the past.

5.1 ULTIMATE CAUSATION

Ultimate causes are evolutionary ones—the social or environmental demands in the distant past that favored some traits over others because those selected traits aided survival and reproductive success. Ultimate causes account for how a trait came to be; they do *not* account for its current existence. The forces in antiquity that contributed to the persistence of a trait may no longer be relevant and may not now support the trait.

During the past 10,000 years, however, humans have increasingly lived and thrived in less natural, more constructed environments, giving emphasis to the social evolution of traits (Symons, 1979/1987). The explication of current or immediate causes of a trait is a matter of proximate causation, which is taken up in the second section of this chapter.

5.1.1 Evolutionary Foundations: From Sex Drive to Socialized Attraction

Modern humans (*Homo sapiens sapiens*) appeared in great numbers only some 40,000 years ago. Our *Homo sapien* ancestors date back to more than 200,000 years ago, and the earliest representatives of the *Homo* species appeared some 2 million years ago (Geary, 1998). While human lineage prior to 2 million years ago is controversial, differential mating strategies had already produced physical and behavioral differences between the sexes by this time (Geary, 1998).

The tasks of survival and raising offspring to reproductive age are major forces behind natural and sexual selection. Individuals who did not survive to adulthood, or who did not have children who themselves had children, did not pass on their genes, although some genes survived if an individual's siblings had children. Among early hominids, genetic survival depended on living long enough to have sexual contact with someone of the other sex—contact that resulted in conception—and survival of offspring to reproductive age. This feat required no small amount of effort. The natural environment was harsh and filled with various hostile threats from animals, insects, and disease, as well as threat from kith and kin and marauding bands. Genetic survival depended upon possession of traits that facilitated individual survival and reproductive success (Buss, 1998). Today we are living representatives of our ancestors' success. In other words, many current physical and behavioral differences between the sexes are products of past fitness over a host of inhospitable factors in the natural environment. It should also be noted that long life and many children were also a matter of opportunity and luck for our ancestors. The best genes were no insurance from fatal accidents. However, our discussion is confined to nonrandom forces that affected the evolution of human sexual attraction.

Sexual contacts among our hominid ancestors were not random events; reproduction could not occur without female-male sexual activity and at least male orgasm. Although sexual activity did not always include conception, reproduction was intimately linked to female-male sexual activity. Not until this century with the availability of inexpensive and reliable contraception and the development of effective *in vitro* fertilization procedures has the link between sexual activity and reproduction been broken.[1] To state the case clearly, the lack of female-male sexual contact, unsuccessful conception, or early death of offspring meant that an individual's genes were not passed on to the next generation. However, *gene survival was not a matter of happenstance.* If reproduction by animals, particularly hominids, was ever a random event, that time has long passed because those organisms that possessed traits to promote sexual contact had more offspring than did organisms without those traits. Thus, *traits that promoted sexual contact survived.*

For many animals, sexual behavior is regulated almost reflexively by internal and external mechanisms, including circulating serohormones, signal detection of receptive partners, and previous sexual experience that urges sexual contact when particular social cues are present. Primates and nonhuman animals, for example, respond to external neurochemical cues that signal sexual availability and mate appropriateness usually through odor detection of hormones in urine or other secretions (Stern & McClintock, 1998). Among nonhumans, the hypothalamus is responsible for regulating the level of circulating sex hormones and sexual readiness (Gorski *et al.*, 1978; LeVay, 1993). For this reason, biological researchers like LeVay assumed that the hypothalamus played a similar role in human sexuality.

Evidence of ancient sexual signaling in contemporary humans can be found in responses to sweat or saliva. As discussed in chapter 4, women who live in close space such as university dormitories synchronize menstruation (McClintock, 1971, 1984) and contact with sweat from young females shortens or lengthens a woman's own cycle (Stern & McClintock, 1998). Women also respond favorably to sweat odors of men who are immunologically dissimilar to them (Wedekind *et al.*, 1995). In addition, applications of distilled female sweat or synthesized male pheromones appear to increase the frequency of female-male sexual activity or at least the increase the perceived sexual attractiveness of recipients of the extract (Cutler, 1999; Cutler, Friedmann, & McCoy, 1998). In short, features of biologically regulated mechanisms that facilitate and motivate sexual contact are evident among contemporary humans. It is likely that sexual drive among premodern humans was even more closely tied to biology than it is today.

Early hominids lived in small kin-groups in loose sex-segregated bands, occasionally encountering other bands. As these creatures formed more socially elaborate relationships and communities and connections to other communities, demands for skills to detect and express sexual interest and receptivity challenged and transformed biological sex drive. As survival in the kin-group or community became more dependent upon successfully navigating the increasingly complex social environment, the biological drive for sexual contact acquired greater social meaning and became less direct—a shadow of its past form and a new more pervasive phenomena, adapted and incorporated into a variety of relationships. The socialization of the biological sex drive is likely to have coincided with greater specialization of the hominid brain and its volumetric increase (Brothers, 1997). A marked jump in estimated brain volume—approximately a 150% increase—is detected among the earliest humans (*Homo habilis*) 2 million-years ago compared to their nearest relatives (*Australopithecus africanus*), an increase not attributable to increase in physical size (Geary, 1998; McHenry, 1994).

Equally dramatic increases in brain volume are evident again with the appearance of *Homo erectus* and yet again with *Homo sapiens* (Geary, 1998).

At about the same time that the social world acquired greater significance for hominids, specialized cognitive-social competencies developed and were exploited (Brothers, 1997). As life became more structured and defined by social roles, meanings and context, individual and genetic survival through sexual contact increasingly necessitated language skills, abstract concepts, acculturation, emotional relationships, and the skillful negotiation of social interactions. (This, incidentally, was also the beginning of an enduring human culture that was transmitted socially and orally from one generation to the next.) In the largely social world of constructed roles and abstract ideas, individual survival and reproductive success demanded quick and accurate interpretation of social information, cooperation, adjustment to communal attitudes, and manipulation of social meaning and position. The need to detect, interpret, and express complex layers of social signals and motivations increased many-fold. Those individuals who could meet this demand achieved sexual contact and passed on their inherited traits to offspring. The increased need for social communication, rapid information processing, and language skills, as well as an increased living range and tool-making abilities, probably account for the dramatic increase in estimated brain volume at particular points in human evolution.

Language is intimately linked to social communication, and reliance on language and social discourse facilitated the emergence of self-consciousness among early humans (Brothers, 1997). In referring to the emotions and motivations of others in relationship to oneself, early humans became autonomous psychological beings in possession of their own desires and feelings. Individual survival and successful sexual contact rested on the accurate detection and interpretation of another's motives and feelings, as well as manipulation of or cooperation with those motives and feelings. Thus, identifying one's own sexual motives and feelings and accurate interpretation of the sexual motivations of others became important elements in the game of reproductive success.

No doubt the increased importance of processing social stimuli resulted in changes and adaptations in the hominid brain (Brothers, 1997). Compared with other primates, and most animals, humans have a very large brain-to-body-weight ratio, as well as larger frontal lobes, amygdaloid bodies, cerebral cortex, and association areas. Not coincidentally, these areas are responsible for processing social stimuli. What is more, structures in this general region—the nucleus accumbens, prefrontal cortex, basolateral amygdala, and hippocampus—are implicated as the so-called pleasure circuitry of the brain (Gallagher & Chiba, 1996; Robbins & Everitt, 1996). In other words, social significance and emotions are closely related in the cerebral cortex.

Brothers (1997) argues from neurological evidence that the human brain is specialized to process social stimuli and produce cognitive symbols of the world. In fact, she claims that heavy reliance on social information among early hominids created a brain *bias* toward the detection and interpretation of social phenomena and the construction of mental representations. Brothers speculates that reliance on oral communication among the earliest humans facilitated the ability to experience others as psychological beings, become self-conscious, and create complex feeling states or mood. This last effect is particularly significant, according to Brothers. By attaching feelings—anger, fear, guilt, remorse, depression, joy, altruism, envy, empathy, acceptance, affection, attraction, and love—to social experiences, the quality and significance of events were radically enhanced, and the range of possible meanings became virtually limitless. Verbal communication may be packed with abstract concepts, complex feelings, and layers of motives. Once this happened, it was not only important to detect someone else's true emotional state and motives, but the individual needed also to mask or guard against having their feelings and intentions known too soon. Similar to modern times, deception and counter-deception became common elements of social and sexual negotiation, verbal and otherwise.

During this dynamic evolutionary period in which biological sex drive was increasingly socialized, it is likely that sexual attraction—more precisely, the sex to which one is attracted—emerged independent of sexual motivation, as *a special case of social affiliation.* Sexual attraction to and a sexual relationship with the other sex were forms of affiliation that differed in degree and kind from other affiliate relationships. Sexual relationships varied in emotional and physical intimacy, tenderness, passion, and commitment. What is more, it is likely that particular sexual relationships acquired social significance at this time. Consistent with the values of the developing culture and the social positions of participants, sexual attractions and relationships could be considered appropriate or inappropriate, special or casual, respectful or disdainful, right or wrong, good or bad, and moral or immoral. *The convergence of social significance, context, and role with biological sex drive among larger-brained hominids resulted in the emergence of sexual attraction.*

The birth of sexual attraction and its close association with emotion states are what made eroticism and romantic love possible. Eroticism is an intense intimate, affectionate nonkin attraction atop an implicit sexual interest, although this interest may not be explicitly sexual. Eroticism can be thought of as a general low-level sexual attraction. *Eroticism may be the glue that binds nonkin affiliate relationships.*

Sexual attractions and eroticism serve different purposes for each sex. For both sexes, the ultimate function of other-sex erotic attraction is the facilitation of sexual contact and conception. In loose bands of small sex-segregated kin-groups, other-sex erotic attraction functioned to bring a male and female together long enough and often enough to conceive a child, despite their lack of familiarity with and probable mistrust of each other. The obstacles to successful sexual contact between males and females were many; in these bands of early humans, individuals probably spent most of their time with same sex members. Girls and women gathered food, cooked, and cared for young children; male youth and men were together hunting, scouting, planning, competing, or fighting. Living with same sex members of the community was the norm for early hominids (Geary, 1998). Life was not like the contemporary suburban family in which husband and wife spend a great deal of time together and share equally in the work load and decision making. For hominids, other-sex erotic attraction is likely to have motivated initial social contact with the other-sex, despite social barriers. Through repeated contacts, trust and agreement may have been established or at least resistance was reduced, building up to sexual contact. Of course, sexual attraction was not present in every case of sexual activity. In circumstances of restricted access to sexual partners, convenience, lack of consequences, or sexual assault, sexual attraction may even have been low or absent.

Other-sex erotic attraction is an emotional tendency or predisposition—not an automatic reflex—that is modified by social factors. Other-sex erotic attraction is modified by relatedness to the targeted individual, degree to which sustenance needs are met, availability of high quality sexual partners, as well as by cultural definitions of attractiveness and social status.

Among early hominid communities, no doubt each sex exploited other-sex erotic attraction to advantage their reproductive success. Consistent with male reproductive strategies, other-sex erotic attraction probably fueled the motivation to mate with many female partners. Consistent with female reproductive strategies, however, other-sex erotic attraction likely played a significant role in promoting a long-term emotional relationship with a mate. For females, the long-term presence of a cooperative male improved the likelihood of conception, contributed higher quality food and additional resources to care of offspring, and provided physical protection to self and children. Although hominid males may not have been keen on the idea of being attached to one or even a few females, long-term sexual relationships had certain advantages for them too, strengthening alliances with nonkin groups and making sexual activity relatively easy and available. Other-sex erotic attraction probably helped males and females overcome many social obstacles and maintain cooperative intersex relationships even for a short time.

Current data supports this supposition. In general, evolutionists note that men pursue a mixed reproductive strategy of seeking brief, low-investment relationships with some women and long-term high investments with others (Buss & Schmitt, 1993; Symons, 1979/1987). In addition, men are generally attracted to bright, cooperative, young, physically attractive, symmetrically shaped women who will be sexually faithful. Men are twice as likely as women to have sexual fantasies at least once a day, four times more likely to focus on the physical attributes of their imagined partner, and three times more likely to fantasize about having sex with someone with whom they are not involved (Geary, 1998). Women are more likely than men to fantasize about sex with their current partner and focus on their partner's personal or emotional traits. Women are more likely than men to avoid casual sexual relationships and are more selective in their partners. Women in general are also attracted to intelligent, resourceful, older, tall, athletic, physically attractive, symmetrically shaped men who will be emotionally faithful. On the whole, women pursue a reproductive strategy of selectively seeking a few high-quality males who will invest resources in the relationship and in their offspring.

To summarize, sexual attraction developed from the socialization of biological sex drive, as a result of hominid life in an increasingly socially constructed world. The emergence of sexual attraction probably coincided with expansive reliance on language and increased brain volume, particularly in areas of the brain that are responsible for language and social information processing. Thus, *other-sex erotic attraction is an adaptation because it facilitated female-male sexual contact and increased the likelihood of producing offspring who would survive to reproductive age.* Because early hominid life was mostly lived in same sex groups, other-sex erotic attraction motivated social contact in spite of inherent mistrust and lack of familiarity with the other sex. Furthermore, each sex employed other-sex eroticism to benefit their own reproductive strategy. For males, other-sex eroticism took the form of explicit sexual desire, which motivated sexual activity; for females, other-sex eroticism was expressed as emotional attachment or romantic sentiment, which helped strengthen commitment to and by a selected mate.

5.1.2 Social Function of Same-Sex Eroticism

As stated at the beginning of this book, the linchpin of a good general theory of human sexual attraction rests on a logical, conceptually consistent explanation of persistent same-sex erotic attraction. This text proposes that *early humans adapted same-sex eroticism to manage intrasex conflict and facilitate same-sex alliances.* This supposition rests on a key assumption—that *sexual attraction is a heterozygous trait.* The two versions of the sexual attraction trait are other-sex eroticism and same-sex eroticism. As discussed in chapter 4, heterozygous traits are the optimal genotypes when the natural environment is unstable—and I would add here, the social environment, as well.   In unstable environments, mixed genotypes maximize genetic variability and adaptability. In other words, the individual with both versions of the sexual attraction trait *should* have a selective advantage over individuals with only single versions of the trait, given an unstable, changing environment.   Indeed, the natural and social environments of our early hominid ancestors were harsh, hostile, dangerous, ever-changing, and unforgiving. Therefore, it is likely that heterozygous genotypes for sexual attraction conferred selective advantages to their hosts compared with homozygous genotypes. Even so, both genotypes should persistent over time.

Before addressing further the question of the reproductive superiority of heterozygous over homozygous genotypes for sexual attraction, we must stop for a moment and discuss how same-sex eroticism was adapted to facilitate intrasex relationships.

5.1.2.1 Same-Sex Female Eroticism

In most human societies, women spend the majority of their lives in sex-segregated nonkin groups (Geary, 1998). In many cultures, prior to marriage, women live in small close knit groups of female kin. After marriage—and marriage in many cultures is mandatory—women leave their kin group to live with their husband's female relatives. Although women may have some contact with kin after marriage, for the most part new brides enter into an established female social hierarchy and the majority of their time is spent within this new group of female nonkin. Negotiating the crags and crevices of this social group to establish position and gain a share of material and nutritional resources requires considerable skill.

Comparative and cultural anthropological data support the contention that premodern human female groups established social hierarchies, although somewhat differently and to a less extent than did male groups (Geary, 1998; Symons, 1979/1987). Females groups have an *alpha* female, but all-female groups are less likely than males to use physical aggression or threat to establish dominance and regulate members' behavior, which is not to say that females never use physically aggressive tactics. Female groups are more likely than males to employ shunning, ridicule, and manipulation of emotional relationships to establish social position.

The presence of new members in an established social group, such as a nonkin or distant-kin bride, alters group dynamics and challenges each member's social position. In spite of community approval of a marriage, a newly-wed girl who enters a long formed kin group of her husband's female relatives may face hostility, a lack of support, or manipulation as social positions are shifted and reestablished. In order to survive, one of the first tasks facing a new bride is to form cooperative relationships with powerful women in the group, similar to the kinds of cooperative, protective relationships that she enjoyed in her own kin-group. Cooperative relationships enhance survival (Trivers, 1985), and in ancient societies cooperative social alliances were critical to survival. Women who formed a devoted and trusting alliance with a same sex group member, especially a powerful group member, could improve their access to high-quality food and more successfully fend off attacks by single females or factions of females. In addition, once pregnant, women faced a considerable risk. Pregnant women were less able to obtain quality resources without assistance and more vulnerable to attack. Because male partners may not reliably provide assistance or protection, establishing loyal same sex help-mates was critical for women to survive pregnancy and childbirth and assist with child care, especially infant care. The importance of interpersonal cooperation among hominid females cannot be over-estimated. Women who effectively negotiated established same sex social networks and developed significant cooperative relationships with one or more members of the group enhanced their chances of survival and enhanced the probability that their children would survive to adulthood. In other words, women who possessed traits that facilitated the formation of close cooperative same sex alliances had a selective advantage over women who lacked such traits.

By cooperative erotic alliances between women, I mean more than a mutual agreement to share work load, more than an uneasy acquaintanceship, and more than friendship. Too often cultural anthropologists have homogenized "friendship" to mean a general familiar cooperative relationship. No doubt this kind of friendship played an important role in surviving intrasex tensions and conflict because having good relations with most members lessened the likelihood of personal attack. Still, friendly relationships were a matter of degree. What was important for survival in same sex female groups were special emotional alliances and coalitions because these relationships strengthened one's position in the group and enhanced control over available resources. *Special cooperative erotic alliances were trusting, intimate, affectionate, long-term, and mutually rewarding*; they were not uneasy, distrustful social games between potential rivals. Affectionate, loyal alliances were pleasurable. Such relationships were evident by an intimate emotional quality similar to that found among inseparable best friends, close family members, mentor and mentored, and romantic lovers.

Special erotic alliances among hominid females were not necessarily sexual relationships, although perhaps many were. These alliances were *erotic* in the sense of an intense affectionate intimacy that excluded others. In fact, the expression of affection through sexual behavior may have strengthened emotional bonds between female pairs. Same-sex sexuality posed little risk to reproductive success, as long as some sexual partners were male. In general, hominid women who lived in sex-segregated societies had few opportunities to engage in a variety of sexual relationships compared with men. Yet same-sex partners were available, and same-sex erotic activity probably went unnoticed or was ignored by men. Among intimate female dyads, explicit sexual behavior was less important than erotic affection, although perhaps not less important to contemporary readers from our highly sexualized Western culture.

In short, *among early human females, same-sex eroticism functioned to promote emotionally significant relationships in order to enhance social position, reduce intrasex conflict, acquire quality resources, and ultimately to improve reproductive success.* Thus, *same-sex eroticism is an adaptive trait.* Early hominid women who possessed this trait were better able to form loyal intimate alliances and coalitions, gain control over available material and nutrition resources, ensure help and protection during vulnerable periods such as childbirth, and obtain assistance with child care. Ultimately, having emotionally significant same-sex relationships improved the chances that a woman's offspring would survive to reproductive age.

As noted in previous chapters, long-term intimate transgenerational relationships are quite common among women in sex-segregated societies and in polygynous households (Blackwood, 1993). The persistence of these relationships supports the notion that they provide practical benefits to women. For example, among the Lesotho of southern Africa, Mummy-Baby relationships between an older women and a young unmarried girl function as social and sexual "finishing schools," as well as sources of emotional support after marriage (Gay, 1986). Likewise, Mombasan women form special same sex erotic relationships that provide emotional support, even after marriage (Shepherd, 1987). Romantic friendships between married women in late 19th-century Western culture may be a variation of this practice (Faderman, 1981/1998).

Same-sex eroticism is probably the emotional accelerant that fuels intimate affectionate relationships between women. More generally, low level eroticism binds same sex social networks and maintains distrust of the other sex.

5.1.2.2 Same-Sex Male Eroticism

In most human societies, males spend the majority of their time with other males, not with females (Geary, 1998). Contact with females is limited and may only occur under circumscribed conditions. In their same sex group, males more than females establish rigid social hierarchies that are closely regulated through physically aggressive competitive play and fighting. In fact, most male-male activities across cultures are competitive and are designed to produce winners and losers. Competition for sexual partners is just one form of male-male competition. Among males, natural and sexual selection has favored intrasex aggression, competitiveness, social hierarchies, physical strength, and a larger size, as well as coordinated group activities such as hunting and fighting (Geary, 1998).

Unlike females, males in early hominid societies probably remained with their kin-group most of their lives and associated more often with nonkin individuals, who were likely to be other males. Consequently, established male hierarchies may have been less vulnerable to instability caused by the presence of new members. Still, all evidence suggests that for hominids intrasex male conflicts were always close to the surface. Against this backdrop of intense competition and aggression, males who lacked alliances or protective kin were vulnerable and at risk of being dominated or killed by powerful males. Again, anthropological evidence suggests that cooperative relationships enhanced survival (Trivers, 1985), and such relationships were critical within same sex groups. In ancient societies, strong social alliances were necessary for survival and ultimately for reproductive success.

Against aggressive individual males or coalitions of males, a trusted alliance with another served to strengthen protective defenses, reduce the threat of hostile attack from opposing factions, improve strategic offenses, facilitate access to quality material and nutritional resources, and aid access to quality sexual partners. For hominid males, loyal same-sex alliances and coalitions improved coordination of efforts during hunting and when defending against enemy forces. Especially, during the latter, one needed a loyal ally who was willing to risk his life if necessary to resist hostile attack. What is more, dominant males and powerful male coalitions had more choice in sexual partner than low-ranking males who may have had no choice at all. Finally, close affectionate male alliances, especially between older and younger males, had political benefits within the community; powerful individuals with social backing or influential factions had more say in community decisions that directly benefited them. Young men gained by close emotional ties to older men.

Again, the term friendship poorly describes these close affectionate alliances, and previous writers have used this term too casually. "Friend" is an old term that has various meanings because friendships vary in affectionate intimacy and loyalty. In the ancient world, friends were very important (Boswell, 1980; Brooten, 1996), and quality of friendship was critical to survival. Furthermore, from my extensive survey of same-sex erotic relationships across cultures, the terms most often employed to characterize same-sex erotic alliances between men are *friend* and *brother*. Obviously, in this context, friend means something more than a casual acquaintanceship. Given the likelihood of the term's casual misinterpretation, I avoid using the term *friend* to describe same-sex erotic alliances.

Special erotic alliances between early male hominids were emotionally charged and probably few in number. Because of the investment involved, it is unlikely that most men maintained more than one or two such relationship, although particular men of high social status may have had many erotic alliances. Male-male erotic alliances were trusting, respectful, mutually rewarding, pleasurable, and intimate to the exclusion of others. Devoted alliances were probably characterized by oaths of fidelity, passion, and violent jealousy. No doubt, the fierceness of male life among early hominids spilled over into erotic relationships.

Special alliances among early male hominids were erotic and possibly sexual. Given the structure of most ancient human societies, it is quite probable that hominid men found emotional significance and passion in their intimate relationships with other men and not in their relationships with women, which were likely to be strained by an uneasy social inequity and distrust. Men were equals to men, or at least they were more equal than women were to men; this social disparity among hominid sexes is also true for most human cultures until recently. Men were also near; men spent most of their time with other men. And men, being more explicitly sexual and tending to express affection in sexual terms, probably expressed their feelings and commitment to a partner through sexual acts. Indeed, sexual activity between male partners may have enhanced their emotional bond and marked their relationship as special to each other and to other members of the male community. It is important to note here that until early modern Western history, intimate physical behavior between males for pleasure or orgasm was not itself a problem, as long as the activity did not exclude sexual behaviors with females and was consistent with social roles or age norms. Furthermore, so long as sexual activity was not exclusive to males, sexual intimacy between males did not risk reproductive success. A close erotic alliance between men could enhance each other's social success and survival. Lastly, in many early human societies, erotic alliances between men may have been the rare social opportunity for expression of passionate love. Until late modern times, romantic love was not pertinent to male-female relationships or marriage. Ancient Greek men, for example, were not expected to fall in love with their wives, although they were expected to love them (Boswell, 1980; Dover, 1978/1989). According to Greek (male) values, deep romantic love was only possible between men because women were social inferiors. Certainly, marriage did not prohibit Greek men from expressing romantic love to males. Also, recall Tokugawa Japan where the idea of passionate romantic love that initially characterized only *nanshoku* brotherhood bonds was eventually adopted as requisite to marriage during in the late 17th century (Leupp, 1995). Here, romantic love was a very late addition to marriage—an element that originated first among male-male erotic relationships.

Thus, *among early human males, same-sex eroticism functioned to support special erotic alliances that enhanced social position, managed intrasex competition, enhanced the acquisition of high quality resources such as female sexual partners, and ultimately improved reproductive success. For men, same-sex eroticism is an adaptive trait.* Long-term and even lifelong special male erotic alliances are evident across cultures between initiate/mentor, beloved/lover, apprentice/warrior, and even occasionally between social equals (Boswell, 1980, 1994; Dover, 1978/1989; Greenberg, 1988). The persistence of similar erotic alliances across cultures suggests that these relationships have significant emotional, social, and political benefits for participants.

What is more, a low level eroticism infuses male social networks and hierarchies. On a group level, same-sex eroticism is the adhesive that binds "old boy" networks, political patriarchies, sports teams, and military units; it is the loyalty and devotion that men feel for close male partnerships and male-male activities, even though in general men view each other as competitors. Evidence of same-sex eroticism among groups of men is evident in the rancor and sense of sacred violation voiced by in-group members when someone of the other sex intrudes on their space. Politics, the military, private clubs, and the gymnasium are traditional male activities and space. Although in today's world there are few exclusive male spaces, locker rooms are still male-space that offer good examples of in-group reinforcement. When men are alone together, they often engage in sexually aggressive, misogynist, and homophobic banter that is designed to challenge or to include (Pronger, 1990). In locker rooms, the extent of misogynist and homophobic remarks is often associated with a sense of sexual vulnerability or objectification among males who are in stages of undress and exposed. However, these remarks, rather than advocate violence or explicit sexual objectification, confirm the heterosexuality and masculinity of participants, assert their in-group identification, and manipulate present roles and perceptions to gain social position. The frequent assertion of "heterosexual" masculinity among men in male-space prevents the situation and speakers from being labeled *queer* (Hopkins, 1992; Pronger, 1990). At the same time such sexually charged banter promotes an erotic tension between members in the group.

All-male groups such as sports teams, military units, and paramilitary groups tend to advocate the formation of close, cooperative, trusting relationships and even intimate pair-bonding so that members or partners can "count on" each other to work together and defend the other in the face of hostile forces, even to the point of risking one's life. The Sacred Band of Thebes was actually an extreme example of this idea. Theban warrior-lovers were chosen for the very reason that a lover would surely not disgrace or forsake his beloved in battle. Ironically, today, militaries worry that same-sex eroticism will do just the opposite and reduce unit cohesion and efficiency. Pervasive homonegative fear that close male relationships will be sexualized and violent reactions in response to that vocal fear support the idea that contemporary male alliances are or can be erotic. Homonegative anxiety about the sexualization of male relationships was at the heart of U.S. military policies to exclude openly "gay" men and "lesbians" from serving. During the 1993 U.S. Senate hearings on gays in the military led by Senator Sam Nunn (D-GA), speakers voiced worries that in lifting the ban "heterosexual" men would be viewed as sexual objects by other men and group cohesion would be corrupted by the presence of men who sexualized male relationships (Kauth & Landis, 1996; "Nunn offers," 1993).

Same-sex eroticism is also evident in manhood initiations. Male initiation rituals are common among sex-segregated, all-male communities, as previously noted. Hazing rituals are also a form of manhood initiation. Manhood rituals often require initiates to engage in low-status, "feminine," humiliating, unpleasant, painful, sexual or erotic, repetitious, or dangerous behaviors during a secret ceremony, while they are partially or completely unclothed. Lengthy male initiation rituals among the Sambia of Melanesia are one example. For years Sambian boys fellate adult male bachelors to orgasm in order to obtain male semen, become fertile, and grow to be men (Herdt, 1981; Herdt & Stoller, 1989). Given the belief that boys must obtain semen from adult men, the Sambia have chosen a particularly sexual method for exchanging this substance. Fellatio is not the only way to get semen, if that is the only goal. Although Gilbert Herdt, the leading chronicler of Sambian life, downplays the sexual aspect of boy-initiation rituals, his side comments and footnoted observations undercut his argument. Indeed the intimate sexuality of manhood initiation among the Sambia seems to be central to male-group identification and establishment of strong male alliances. Similar outcomes are evident in other manhood initiation rituals that are less explicitly sexual and more homoerotic, such as those in the military, paramilitary groups, and fraternities. At least one objective of these rituals is to create a close bond among men.

To summarize these last two sections on the social function of same-sex eroticism, sexual attraction comes in at least two forms, and a ratio of both forms is likely to confer a selective advantage for individuals over single versions of the trait. Said another way, if sexual attraction is a heterozygous trait, having both same-sex and other-sex eroticism confers a selective advantage over exclusive sexual eroticism. Given the natural and social conditions of early hominid life, it is likely that sexual attractions were socialized and adapted by the sexes for different purposes. The ultimate function of other-sex erotic attraction for both sexes was the facilitation of sexual contact and conception, regardless of what other benefits were associated with this kind of attraction. However, in general, same-sex eroticism probably functioned to reduce intrasex conflict and promote emotionally significant alliances. For females, erotic same-sex alliances defended against hostility from nonkin females, provided support during critical times in life such as late stage pregnancy and care of a newborn, and increased the likelihood of obtaining quality food. For males, erotic same-sex alliances reduced intrasex aggression, aided individual survival, enhanced social status and power, and provided greater access to high-quality resources such as food and female sexual partners. Although sexual behavior may have cemented or signified the special closeness of a same-sex relationship, its presence and frequency or its absence was secondary to the emotional eroticism that maintained such relationships. Whatever sexual behaviors occurred within same-sex erotic alliances, these relationships were not sexual in a contemporary sense of the word.

Given this argument, *having a capacity for both same-sex and other-sex erotic attraction permits the widest range of adaptive responses and optimal fitness in terms of reproductive advantage in unstable environments. Thus, on the average, individuals with a ratio of varied sexual attractions were better fit than individuals with an exclusive sexual attraction.* Those who experience an exclusive form of sexual attraction are at least reproductively and perhaps socially disadvantaged, according to this argument.

5.1.2.1 Exclusive Sexual Attraction as a By-Product of Heterozygosity

Exclusive sexual attractions include both exclusive other-sex eroticism and exclusive same-sex eroticism. As a heterozygous trait, some people will have only homozygous versions and an exclusive sexual attraction as a result of genetic variability. Historical records and contemporary surveys suggest that exclusive same-sex erotic attraction is relatively rare, and the new model of sexual attraction asserts that exclusive other-sex erotic attraction is also relatively infrequent. Indeed, recorded examples of exclusive other-sex erotic attraction in the ancient world are difficult to find (Boswell, 1980). In the distant past, exclusive other-sex erotic relationships may have been unusual or just unimportant and not worth mentioning. By contrast, contemporary sex surveys consistently find that the vast majority of men and women identify as "heterosexual" (Janus & Janus, 1993; Kinsey *et al.*, 1948; Kinsey *et al.*, 1953; Laumann *et al.*, 1994). It is worth stating again that a sexual orientation identity is not synonymous with a *predisposed* sexual attraction. For many reasons, already discussed, these phenomena differ. Contemporary sex surveys also point out that sexual behavior and fantasy are not always consistent with reported sexual orientation. The fact that today most people describe themselves as "heterosexual" and engage in predominantly other-sex erotic behaviors—although only a small percentage of men and women may have a biologically disposed exclusive other-sex attraction—is completely consistent with the new model of attraction.

If a heterozygous genotype for sexual attraction represents optimal reproductive fitness, over time more individuals will possess this trait, although not all will. Thus, *the majority of people have the potential for erotic attraction to both sexes, although this potential is not necessarily expressed as explicit sexual behavior.* Eroticism plays a social function that need not be overtly sexual to benefit individuals. However, *given a potential for sexual attraction to both sexes, the cultural environment and social experiences have the defining role in shaping the expression of sexual attraction.* Said another way, given tolerant or hostile social beliefs about same-sex erotic behavior across cultures, the number of so-called "heterosexuals"—those who engage predominantly in other-sex erotic behavior—could wax or wan significantly.

There is a further implication of this idea: *If a heterozygous genotype for sexual attraction represents optimal reproductive fitness, then both exclusive other-sex erotic attraction and exclusive same-sex attraction are less reproductively advantageous.* Readers will probably have no argument that exclusive same-sex erotic attraction is no reproductive advantage; having little or no erotic desire for the other-sex makes conception through sexual contact less probable and less efficient than for those who desire other-sex erotic contact. Yet the difference in reproductive advantage between individuals with heterozygous genotypes for undifferentiated erotic attraction and those with homozygous genotypes for exclusive other-sex attraction is equally dramatic. Characteristics associated with attraction genes are not limited to prompting sexual contact. In fact, in this new model, prompting sexual contact is one of the cruder functions of sexual attraction genes. *Although sexual contact and successful conception is critical to reproductive success, a host of associated sociosexual characteristics are no less important such as social and communication skills related to managing intrasex conflict and forming cooperative erotic alliances.* Male and female hominids who had insufficient socioerotic interest or motivation to initiate and maintain significant same-sex relationships no doubt found themselves marginalized, dominated, harassed, and vulnerable to personal attack; such individuals would not have been in good positions to exercise an other-sex erotic interest.

In fact, without a balanced interest in same-sex relationships, the hominid community could easily perceive a strong exclusive other-sex erotic interest as a threat to established social order.

The new model predicts that early humans with a ratio of sexual attractions had a greater likelihood of successfully negotiating a complex, changing, and hostile social environment and passing on their genes to surviving offspring than did individuals who had other-sex erotic attraction but lacked same-sex socioerotic traits. In other words, ancient hominids with a ratio of sexual attractions should have had more varied social relationships and produced more children who lived to adulthood than males and females with only an exclusive other-sex erotic attraction. This is not to say that males or females with an exclusive other-sex erotic attraction could *not* form close alliances or friendships with members of the same-sex. Rather, I argue that hominids with varied sexual attractions established close same-sex alliances, achieved other-sex erotic contact, and produced surviving offspring more frequently and successfully than did hominids with only an exclusive other-sex erotic attraction.

There is another implication of this model: Given the optimal fitness of heterozygous genotypes for undifferentiated sexual attraction, exclusive other-sex eroticism and exclusive same-sex eroticism are similar in several respects. Neither form of exclusive sexual attraction is more common or more privileged than is the other, and neither form of exclusive attraction is optimal in a reproductive sense. In fact, both exclusive forms of sexual attraction have their own inherent disadvantages, reproductively speaking.

5.2 PROXIMATE CAUSATION

Proximate causes are near or immediate ones—in this case, the neurohormonal-social experiential mechanisms that account for the development of varied sexual attractions. Evolutionary pressures on sexual attraction influenced change through genetic and neurohormonal mechanisms and cognitive-affective processes, although the forces that support varied sexual attractions today may not be the same ones that supported development of varied sexual attractions in ancient times. In examining proximate causes of sexual attraction, we focus on explanations within our experience rather than on ancient or distant explanations.

This section presents a new biopsychosocial model of varied sexual attractions, which is contrasted later with a current popular model. The process of moving from a biological predisposition to an erotic trait to sexual orientation is also discussed. Predisposed information processing systems are credited with contributing to erotic association and the consolidation of reinforced experiences into stable erotic templates and sexual orientation. The new model of sexual attraction assumes that, like Nature, variability and diversity are common and advantageous.

5.2.1 A Biological Model of Sexual Attraction

In general, scientists have assumed that more than one gene is involved in sexual attraction. Unfortunately, perhaps for convenience, scientists (Hamer & Copeland, 1994) sometimes refer to sexual attraction genes in the singular, giving the impression that there is a single "gay gene." Equally unfortunate, careless language and pressure to condense highly technical information into short media sound bites has fostered the misleading concept and term *gay gene* in popular culture. However, it is very unlikely that a complex trait such as sexual attraction is directed or influenced by a single gene. Sexual attraction traits are more likely to be the product of multiple genes.

The biological model described here assumes that sexual attraction is polygenetic and that the trait comes in two forms—other-sex erotic attraction and same-sex erotic attraction. For evolutionary reasons, other-sex eroticism is the dominant form of the trait. A polygenetic model of sexual attraction permits a complex combination of genes and the expression of degrees of both dominant and recessive forms of the trait.

Because sex, sexual drive, and sexual attraction are closely associated, genes related to sexual attraction are likely to reside on the sex chromosomes. Again, genes exert their influence by guiding enzyme processes and hormone production and exposure. *During fetal development, genes influence the timing, duration, and ratio of sex hormones to which the individual is exposed, thereby altering the receptivity of neural structures and establishing their function.* Hormone exposure is not a single absolute event. Through *patterns* of sex hormone exposure during critical periods, neural structures are predisposed toward particular functions, stimuli, and associations. In this way, *genes associated with sexual attraction create erotic predispositions in how key neural systems receive and process particular sex-related stimuli.*

Creating an erotic predisposition in neural systems is analogous to drawing lines on a blank piece of paper. The lines order and limit how information best fits on the paper. Depending on the direction of lines, information flows from left to right or from top to bottom, or perhaps even diagonally. Similarly, a pattern of hormone exposure may sensitize key neural structures to certain sex-related stimuli and push erotic processing in a particular direction. To extend the lines-on-paper analogy, hormone exposure creates a grid or structural pattern on neural systems that favors a particular information organization and type of emotional significance over others. Thus, *a sex-specific erotic predisposition increases the likelihood that stimuli associated with the preferred sex are attributed sexual meaning.* Likewise, stimuli associated with the nonpreferred sex are not be assigned sexual meaning.

Creating an undifferentiated erotic predisposition is akin to superimposing both horizontal and vertical lines on a sheet of paper—as a faint grid in which neither horizontal nor vertical lines dominate—that permits a variety of organizational structures. Neural systems with an undifferentiated erotic predisposition can make erotic attributions to a wide range of stimuli associated with males or females. Given different ratios of attraction, as well as cultural pressures and various social experiences, undifferentiated eroticism permits the development of many sexualities.

Given variation in the timing, duration, and proportion of fetal sex hormone exposure, at least four distinct erotic predispositions are possible: *undifferentiated eroticism, other-sex eroticism, same-sex eroticism,* and *no sexual attraction.* These different patterns of hormone exposure and resulting erotic predispositions are listed in Table 4. In actuality, undifferentiated erotic types experience specific cultural and social pressures that favor one form of erotic and sexual expression over others, and so they are not undifferentiated for long. The term *undifferentiated erotic predisposition* is meant to illustrate the potential for making erotic and sexual attributions to both male and female stimuli. However, a potential need not be fulfilled or exercised, and erotic attributions need not be expressed as sexual behavior. Thus, cultural conditions may have more influence on sexual expression for undifferentiated erotic types than for sex-specific erotic types. What is more, depending on the culture, undifferentiated erotic types may be difficult to distinguish from individuals with a sex-specific erotic predisposition. However, undifferentiated erotic types reflect heterozygous genotypes for sexual attraction, and sex-specific erotic types reflect homozygous genotypes.

As Table 4 illustrates, variation in ratio, timing, and duration of sex hormone exposure results in a range of erotic types. *Most people are in the undifferentiated erotic categories, which represents a middle range of hormone exposure across dimensions.* A middle range of hormone exposure should permit the greatest variability in erotic feelings and sexual expression. Furthermore, *few individuals have a sex-specific or exclusive erotic predisposition, which represents wider variation in hormone exposure patterns.* A small minority of people may experience no sexual attraction to either males or females—*aneroticism* (Money, 1988)—but desire sexual pleasure, which may take the form of masturbation.

TABLE 4
Typology of Erotic Predispositions by Sex and by Exposure to Prenatal Sex Hormones

	Male (androgen dominant // estrogen)	Female (androgen // estrogen dominant)
	Undifferentiated predisposition	*Undifferentiated predisposition*
Ratio	low-to-high // high	low-to-high // high
Timing	early-to-late // early	early-to-late // early-to-late
Duration	long // long	brief-to-long // long
	Other-sex predisposition	*Other-sex predisposition*
Ratio	high // low	low // high
Timing	early // late	late // early
Duration	long // brief	brief // long
	Same-sex predisposition	*Same-sex predisposition*
Ratio	high // high	high // high
Timing	late // early	early // late
Duration	long // long	long // long
	Absence of sexual attraction	*Absence of sexual attraction*
Ratio	low // low	low // low
Timing	late // late	late // late
Duration	brief // brief	brief // brief

Contemporary sexual orientation labels do *not* correspond to the erotic types illustrated in Table 4. Individuals who identify as "gay" or "straight" may actually have an undifferentiated erotic capacity and a heterozygous genotype. For this reason, research samples of "gay" men and "lesbians" could include a variety of genotypes that are not similar at all.[2] The problem of mixing genotypes in research samples would be greatest for samples of "heterosexuals." The privilege accorded to heterosexuality in the West may strongly influence people with an undifferentiated erotic predisposition to behave sexually, feel, and identify as "heterosexual." However, according to the new model, few so-called "heterosexuals" would have a genotype for exclusive other-sex erotic attraction.

Early versus late exposure to sex hormones alters the course of fetal development (Ellis & Ames, 1987; McConaghy, 1993), but exactly how sexual attraction is affected has not been determined. Similarly, prolonged exposure to sex hormones versus brief exposure is likely to alter development, given different critical periods. Sex differentiation, for instance, occurs on different timetables for males and females. In addition, brain development is a relatively slow process during which time neurons vary in sensitivity and responsiveness to hormones, making structures vulnerable to variable patterns of hormone exposure (LeVay, 1993). Sex hormones, neuromodulators, and neurotransmitters also alter the morphology of neurons by increasing or decreasing the number of synaptic receptors, by "tweaking" the sensitivity of receptors, and by promoting the growth of new dendrites and new neural connections—thus, increasing neural interconnections. For example, the process of forming memories—called *long-term potentiation*—involves physical changes in particular neurons or neural pathways, allowing them to "recognize" and respond quickly to similar stimuli (Frey & Morris, 1997; McKernan & Shinnick-Gallagher, 1997; Rogan *et al.*, 1997). In short, the new model of sexual attraction claims that *variable exposure to fetal sex hormones sensitizes key neural structures to sex-typed stimuli and predisposes sex-specific erotic associations*. Variable "windows of opportunity" in neural sensitivity help explain the mixed pattern of cognitive abilities reported among men and women of different sexual orientations. Neurohormonal processes associated with sexual attraction affect multiple neural structures.

Because sexual attraction is a special case of social behavior, the "social" brain is likely to be the seat of erotic predispositions. The social brain represents a network of structures that include the amygdaloid bodies, hippocampus, prefrontal cortex, and parts of the association cortex (Brothers, 1997). If eroticism overlays social information processing, then the same neuroanatomic structures should be involved. Sexual attraction can then be seen as a *network* of neural structures, rather than a discrete site. Thus, *patterns of fetal sex hormones sensitize and predispose neural groups in the amygdaloid bodies, hippocampus, prefrontal cortex, and the association cortex toward sex-specific stimuli and making particular erotic associations*. The involvement of higher cortical areas in sexual attraction also permits modification of erotic behavior based on social conditions and individual variables. The hypothalamus is likely to play an indirect role in sexual attraction. Consistent with its other functions in regulating basic "drive" states, the human hypothalamus is more likely to urge *indiscriminate* sexual contact. In short, the hypothalamus more probably regulates sexual readiness or "horniness," as opposed to the direction of sexual attraction.

Finally, variability in erotic predispositions is likely to be influenced by the ratio of sex hormones during fetal development. Biological studies of sexual attraction have focused almost exclusively on the role of androgens in the development of (same-sex) sexual attraction and have overlooked the potential influence of estrogens in competing for neural receptors. In some forms and under some circumstances, estrogens mimic androgens (Goy & McEwen, 1980; Harding, 1986; MacLusky & Naftolin, 1981), and data from animal studies suggest convincingly that the presence of both estrogens and androgens are important for healthy sexual functioning (Ogawa *et al.*, 1997). Therefore, *it is likely that the presence of both androgens and estrogens in the brain—in different proportions, for different lengths of time, and in competition for the same receptor sites during variable critical periods of development—permits wide variation in erotic predispositions.*

In brief, the new model of sexual attraction purports that typical fetal hormone exposure lies in an average or middle range. Patterns of hormone exposure that vary by timing, ratio, and duration sensitize neural structures in the social brain and establish erotic predispositions. Most men and women experience an undifferentiated erotic predisposition and have the potential to experience erotic and sexual attraction to both sexes. However, cultural beliefs and social pressures guide sexual feelings and behavior in particular directions. In addition, eroticism is not necessarily sexual and can be expressed as intimate affection. Undifferentiated erotic predispositions are products of heterozygous genotypes for sexual attraction. Sex-specific erotic predispositions are products of homozygous genotypes for attraction and are by-products of typical genetic variation. Variability in the pattern of sex hormone exposure contributes to sex-specific predispositions to experience sexual attraction to one sex. Few people are likely to have sex-specific or exclusive sexual attractions. Given unstable environments, heterozygous genotypes and undifferentiated eroticism optimize reproductive success.

5.2.1.1 From Erotic Predisposition to Sexual Orientation

Learning predispositions have a precedent among infrahumans and are evident among many species (Whalen, 1991). Animals with a learning predisposition are strongly inclined to engage in certain behaviors and make particular associations. Thus, behavior and learning are not random events. For example, pigeons are strongly inclined to peck and will easily learn to peck for a reward (Weinrich, 1987a).[3] Pigeons are not inclined to peck to avoid punishment. Likewise, cats frequently lick themselves, but cats do not easily learn to lick themselves to obtain their release from a box. On the other hand, cats will quickly learn to push or pull something to open the door of a box. Rats have a keen sense of smell and rapidly learn to make complex odor discriminations, although they are poor discriminators of visual cues. However, humans easily learn subtle visual discriminations, but they have greater difficulty learning olfactory discriminations. Learning predispositions reflect the natural selection of traits associated with reproductive fitness for the organism's environment. Said another way, evolution has selected organisms that make particular learned associations. An organism's evolutionary history predisposes it to respond to particular stimuli at particular times and not to other stimuli. Consequently, stimuli are not equivalent; depending on an organism's evolutionary history, some stimuli are more significant than others. The timing of stimuli exposure is also important. As noted in chapter 4, the sequence and timing of key experiences during early development are critical healthy adult functioning among infrahumans (Whalen, 1991).

The acquisition of language by children is an example of a human learning predisposition that also depends on social context and the timing of learning events for efficient development. By the first year of life, neural circuitry and social experience are insufficiently developed to support more than simply naming things (McGraw, 1987). However, by the end of the second year, toddlers are able to link words into phrases, which are the building blocks of complex sentences. After thirty months of age, vocabulary balloons. However, early instruction does not seem to speed language acquisition. Even so, social and verbal stimulation are necessary for efficient language development, and social isolation and cognitive impoverishment obstruct the acquisition of language. In short, language acquisition among human children is a learning predisposition that has a critical period for development. However, efficient language development depends upon appropriate social experiences. The new model of sexual attraction purports that erotic predispositions are similar to learning predispositions.

If so, how do erotic learning predispositions become erotic experiences and eventually a sexual orientation? As noted earlier, centers for processing erotic predispositions are likely to be located in the web of highly interconnected neural circuitry that includes the amygdaloid bodies, hippocampus, prefrontal cortex, and temporal association cortex. Initially, the amygdaloid bodies receive extensive input from advanced sensory centers that process vision, audition, and bodily sensations and changes. These structures evaluate and weigh the significance of stimuli. Thus, the amygdaloid bodies are likely to be responsible for attaching erotic or sexual significance and feelings to words of affection, a gentle touch, a kiss, a longing look or gesture, or an image. Particular stimuli may prompt the hippocampus to initiate the formation of erotic memories or integrate the experience with existing sexual memories. The hippocampus is less directly involved in the expression of erotic predispositions than are the amygdaloid bodies, prefrontal cortex, and association cortex. However, the hippocampus and the frontal cortex are important in the consolidation of individual erotic events into uniform cognitive-emotional patterns or erotic templates that lead eventually to the global experience of sexual orientation.

Determining the meaning of stimuli within a social context and in relationship to past experience is a higher order function involving the frontal cortex and association cortex (Brothers, 1997). Decisions about erotic importance are likely to originate in the prefrontal cortex; memory cascades of previous erotic or sexual events and feelings and connections to similar events and feelings may then be triggered by the association cortex. The structures in the sexual network share information and work together in processing erotic significance. With input regarding social context, previous experience, and pleasurable past associations, the executive functions of the frontal cortex guide behavior.

Like language development, development of sexual attraction requires a certain level of cognitive maturation. There is, however, no evidence that a postnatal critical period exists for sexual attraction that, if missed, results in a *lack* of sexual attraction. *Developmental data suggest that the experience of eroticism requires an awareness of self as separate from others before which time erotic significance is unlikely to occur.* Awareness of self and others as autonomous psychological beings is associated with a level of language and cognitive development that occurs around the ages of 6 or 8 (Damon, 1983). Children younger than 8 years of age are probably unable to experience erotic associations to social interactions or to sex-specific stimuli. Thus, *from about the age of 8, erotic predispositions are likely to play a role in the meaning of children's social experiences and in behavioral decisions, and significant erotic experiences begin to take shape as sexual attraction.*

As discussed in chapter 4, some theorists (Bem, 1996; Byne & Parsons, 1993; McConaghy, 1993) purport that sexual attraction is a function of nonsexual personality traits, especially sex-typical traits. Such traits are thought to contribute to particular social experiences that shape and modify sexual attraction. No doubt, traits associated with sexual attraction are numerous and many are probably nonsexual. However, for reasons noted earlier, the new model of sexual attraction purports that eroticism and sex-typed behavior are parallel but independent dimensions, both related to sex differentiation. Sexual attraction is not tied to masculinity, femininity, passivity, rough-and-tumble play, competitiveness, novelty seeking, reward dependence, harm avoidance, alloparenting, love of sports, or other sex-typed characteristics. Although sex-typical traits are related to sex, these traits have a great deal of intrasex variability. Sex-typed characteristics are *not* closely related to sexual attraction.

Even so, there is merit to McKnight's (1997) contention that genes related to same-sex eroticism may increase non-sex-typed traits such as sexual drive, social and communication skills, charm, and self-confidence. In fact, McKnight views these traits as more typical of "gay" men and desirable by "heterosexual" women in a mate and for her sons. Hypotheses that "gay" men possess characteristics which "heterosexual" women find attractive and that women more often choose male partners who have those characteristics require further investigation. Although there is considerable evidence that "gay" men have sex more frequently and with more partners than do "heterosexual" men, little data are available to support the notion that "gay" and "heterosexual" men differ on non-sex-typed traits. Most research that has examined sexual orientation differences look only at sex-typed traits, where differences are assumed to be. However, the new model of sexual attraction suggests that "gay" and "heterosexual" men possess different non-sex-typed traits. The new model asserts that same-sex eroticism was adapted to facilitate the formation of close same-sex alliances, which reduced intrasex hostility and enhanced the acquisition of higher quality resources.

The point is that *the new model of sexual attraction claims that a prenatally imposed erotic predisposition results in a postnatal bias in social information processing and erotic attachment.* An erotic predisposition sensitizes identification of sex-specific stimuli in the environment and increases the likelihood that pleasurable (romantic/sexual) feelings are attached to these stimuli. An erotic predisposition in children 8 years of age or older may be evident in their heightened emotional attachment to objects, behaviors, and physical characteristics associated with the preferred sex. For example, boys with an erotic predisposition for females may experience a strong emotional reaction and curiosity toward particular female attributes—breasts, genitalia, soft skin, hair color and style, undergarments, emotional support, and mannerisms—and toward young attractive women. Erotic significance is evident in a desire for or appreciation of physical and emotional proximity with attractive or attentive females. Attributes associated with sexual behavior and physical intimacy are more likely to be perceived as erotic than are non-sexy attributes. Girls with a same-sex erotic predisposition may have a similar emotional and physical response to female attributes and young women, desiring a closeness or special relationship with young attentive nonkin females. On the other hand, boys and girls with an undifferentiated erotic predisposition may show similar emotional patterns to both same-sex and other-sex attributes depending on the extent that the social environment supports such responses. By the age of 8, children are well able to recognize the values of their culture regarding sex-appropriate emotional expression, including the acceptability or intolerance of intimate feelings for same-sex nonkin.

In this regard, early social experiences with adult women would be important during the development of other-sex erotic attraction in boys who have an erotic predisposition toward females. Because they are novel and available, characteristics of significant women from a boy's childhood or adolescence are likely to have erotic importance. Of course, these experiences do not need to be with real women; fictional women make powerful erotic models. Boys and girls frequently encounter sexy fictional women in books, movies, television, posters, fashion magazines, and billboards, especially since sexually provocative women are used often to sell products including television programs. Similarly, early experiences with adult men would be important during the development of other-sex eroticism in girls who have a predisposition toward males. Likewise, early same-sex experiences would be important during the development of same-sex eroticism in boys and girls who have this predisposition.

It should be noted here that the absence of same-sex individuals in a child's environment would *not* lead to the absence of same-sex eroticism in children who have such a same-sex predisposition. No published evidence to date suggests that a critical postnatal period exists for erotic learning and that, if the period is missed, a different outcome results. There is also no evidence that the failure to support an early erotic predisposition leads to its atrophy or the development of a contrary erotic disposition. The predisposition is in the neural wiring, and environmental events either support or fail to support an erotic predisposition; environmental events do not change a predisposition.

Once a sex-typed attribute is eroticized, similar stimuli become more salient and attention is directed more often toward those stimuli. Repeated contact with eroticized stimuli further reinforces romantic or sexual feelings associated with the stimuli and reinforces the class that it represents. Over time, single eroticized attributes or events are consolidated into stable, organized cognitive *erotic templates* of experience, perhaps by the hippocampus. Erotic templates facilitate the identification and processing of stimuli that may have erotic significance. Erotic templates in this model— similar to what Money (1988) called *lovemaps*—are organized patterns of sexualized features or situations, such as bleached blonde women with large breasts and long legs. Other cognitive templates of sexualized features might include large women with dark hair, particular body odors, and exaggerated mannerisms. Erotic templates are likely to involve sexual scenarios, such as spanking or humiliation or being seduced, rather than just eroticized physical attributes. Less extreme erotic templates could involve physically intimate scenarios such as bathing or undressing. Thus, *cognitive erotic templates are important in sexual fantasy and play a key role in sensitizing the individual toward sexual readiness and opportunity.*

Boys with an erotic predisposition toward males undergo a similar process in being sensitized to male attributes—muscles, build, body odors, beard, clothing, genitalia, and mannerisms. Models of these attributes are prevalent in the social environment, especially through the media. Boys with a predisposition to eroticize males experience as an intense emotional attachment, curiosity, and physiological arousal to sexy or intimate male attributes. Over time, single eroticized attributes are consolidated as cognitive erotic templates that guide the sexual interpretation of future images, experiences, and social situations.

Boys and girls with undifferentiated erotic predispositions may interpret either or both male and female sex-typed stimuli as erotic, depending on prevailing cultural attitudes and social experiences. Although the mechanism for identifying and processing sex-typed stimuli is similar for males and females, sex differences consistent with differential reproductive strategies influence which sex-specific attributes are eroticized and the form given to eroticism.   In general, males are more likely than females to sexualize objects, physical characteristics, and situations (Bailey *et al.*, 1994). Therefore, regardless of the direction of their erotic predisposition, boys are likely to experience an explicitly sexual eroticism that is more genitally focused. On the other hand, females employ different reproductive strategies and are likely to eroticize personal and emotional characteristics (Geary, 1998; Symons, 1979/1987).   Therefore, regardless of the direction of their erotic predisposition, females are likely to romanticize characteristics of significant individuals or situations. On the average, men will more often experience eroticism as sexual, whereas women will interpret erotic feelings as romantic and emotional. That is not to say, of course, that men experience no romantic feelings or that women have little sexual desire. On the contrary, it is only that romantic and sexual feelings vary by sex in their emphasis.

During puberty, erotic templates are exercised and solidified. Puberty represents sexual maturity and the beginning of sexual life, as well as the actualization of sexual attraction. *By puberty, erotic experiences have already formed cognitive erotic templates and early sexual attraction.* The effect of pubertal sex hormones is to "push" the frequency of making of erotic associations and "turn up" desire for intimate contact with another. During puberty, the hypothalamus plays an instrumental role in sexuality through the regulation of circulating sex hormones and sexual drive. A sexual contact "urge" is the body's way of facilitating reproduction and ultimately genetic survival. Sexual drive is the fuel behind sexual attraction. The hormonal intensification of sexual feelings at puberty is why adolescent boys feel "horny" and adolescent girls develop crushes, consistent with their evolutionary programming. Over time the intensity and novelty of erotic feelings moderate and become somewhat more manageable.

What is more, *during adolescence and young adulthood, cognitive and behavioral rehearsal reinforces frequently used erotic templates and extinguishes alternative ones.*   Reinforcement habituates particular erotic templates and sexual attraction, and *stable erotic templates and attractions become an individual's sexual orientation.* By middle adulthood, variation from routine sexual attraction and behavior is unlikely. Even so, *sexual attraction is not completely fixed or unchanging*; habitual sexual attraction and behavior can be altered by an unusual emotionally significant interpersonal experience, given a supportive social environment.

Cultural beliefs play a major role in shaping the interpretation and experience of erotic predispositions. Because eroticism is a higher cortical function, human sexual attraction is more influenced by cognition than if it were a subcortical hypothalamic event. For this reason, beliefs play a significant role in the meaning and expression of sexual attraction—beliefs about the rightness or wrongness of sexual feelings, the importance of sex in daily life, when and under what conditions sex is appropriate, which sexual acts are taboo, and with whom and which sex one may engage in sexual activities. The effect of cultural beliefs on sexual attraction and behavior is quite evident in cross-cultural examples of male sexual behavior. Cultural tolerance, restriction, or condemnation of same-sex eroticism has contributed to enormous variability in male sexual behavior. For a majority of men in ancient Greece, Rome, and China, and in Tokugawa Japan, same-sex erotic attraction and sexual relationships were not only possible but also expected and even idealized. However, for men in 18th-century Tahiti, native American Indian societies, and the modern era, same-sex eroticism was or is restricted to certain situations or confined to particular gender roles. For men in these cultures, same-sex erotic behavior is forbidden and abhorrent outside of these situations or gender roles. These different sexual cultures do not represent different subspecies of men. Nor is the span of a few hundred years sufficient enough to reduce significantly the population prevalence of a genetic trait for a wide-ranging erotic capacity. Although far less is known about female sexual behavior in most cultures, sufficient evidence supports that notion that women are also capable of a range of sexual feelings and responses that are also shaped, encouraged, or limited by cultural beliefs. Restriction of sexual opportunity through physical confinement and social regulation is one of the primary ways that cultures have controlled female sexual experience.

Again, *given the argument that same-sex eroticism is adaptive, most men and women are likely to possess the capacity for erotic and sexual attraction to both sexes.* The capacity for eroticism to either or both sexes permits a wide range of emotionally significant nonkin relationships, some of which may be sexual.

Contemporary Western culture does not encourage same-sex eroticism, and not surprisingly, most men and women describe themselves as "heterosexual." Reporting same-sex erotic feelings or experiences is not socially acceptable. Even so, modern sex surveys give a small indication that both sexes have a greater erotic capacity than is first suggested by their sexual orientation identification. As noted in chapter 2, a recent probability sample of adult Americans found that nearly 8% of women reported some same-sex erotic attraction or desire and 4.3% had at least one same-sex erotic experience since puberty (Laumann *et al.*, 1994). However, only 1.4% of women in the survey identified as "lesbian" or "bisexual," and 99% identified as "heterosexual." Curiously, more men in the study engaged in same-sex erotic behavior than reported having same-sex erotic desire. Almost 8% of men reported experiencing some same-sex erotic desire, but 9.1% of men had engaged in at least one same-sex erotic experience since puberty. Obviously for men, sexual activity can occur without sexual attraction. However, only 2.8% of men in the sample identified as "gay" or "bisexual," and 97%

identified as "heterosexual." In other words, despite a cultural value *not* to have or report same-sex erotic experiences, a significant number of "heterosexual" men and women not only have same-sex erotic feelings but many have also engaged in explicitly sexual same-sex erotic activities. Given a more supportive social environment, it seems likely that the number of men and women who report same-sex erotic feelings and sexual behavior would increase. At the same time, men and women in these surveys who do not acknowledge experiencing same-sex eroticism are not being false or deceptive; such individuals may be entirely true to their erotic experience, given the cultural context.

To summarize the new model of sexual attraction, multiple genes are associated with eroticism. A heterozygous genotype for sexual attraction represents the optimal fit for reproductive success. Although other-sex eroticism functions to promote sexual contact with the other sex, same-sex eroticism functions to reduce intrasex conflict and competition and increase access to high-quality resources by facilitating the formation of emotionally significant alliances. Most men and women have the capacity for erotic attraction to both sexes; a few men and women, as by-products of genetic variability, have an exclusive or sex-specific erotic predisposition. Guided by genes associated with sexual attraction, a pattern of sex hormone exposure predisposes social information processing in the amygdaloid bodies, prefrontal cortex, and association cortex. An erotic predisposition in information processing systems favors sex-typed stimuli and increases the likelihood that such stimuli are given erotic significance. By middle-childhood, prevailing cultural beliefs and social experiences actively exploit or challenge erotic predispositions, and over time erotic experiences are consolidated into stable erotic scenarios or templates. Habituation reinforces sexual attraction. Erotic experimentation and alternative sexual experiences become less likely by middle-age, unless routine sexuality is disrupted by a significant interpersonal experience and if experimentation is supported by the social environment. In brief, natural and sexual selection has favored humans who possess a wide-ranging erotic potential that, depending upon the strength and proportion of sexual attractions, is shaped to fit the social environment.

5.2.1.2 Comparison to Money's Model

In the scientific community, the most accepted model of sexual orientation at present is one espoused by John Money (1988). Since the introduction of this model, Money's views have remained consistent with the original (Money, 1998). The new model of sexual attraction offers several advantages over Money's model. This section briefly highlights the differences between the two models and the advantages of the new one. A summary of differences between the new model and Money's is provided in Table 5.

First, Money's (1988, 1998) model is largely a proximate theory of homosexuality, not a general theory of human sexual attraction. He focuses on explaining abnormality. Second, Money presents a pathological model of same-sex eroticism, despite the absence of pathology. He accepts that typical Western sexuality—exclusive heterosexuality and gender roles—is the standard of measure. Erotic feelings or sexual behaviors that deviate from this standard are problematic, pathological, and require explanation, in his view. A relatively small portion of Money's model is devoted to the development of other-sex eroticism. Third, he views gender identity, sex-typed behavior, gender role, and sexual attraction as inseparably linked phenomenon, which locks him into a pathological gendered model of attraction when erotic feelings and sexual behavior vary from the standard. The problems inherent in a gendered model are summarized below. And fourth, Money's model contributes to confusion with unhelpful and awkward neologisms such as *gynemimesis, andromimesis, paraphilic strategems, abidance, ycleptance, foredoomance, gendermap, lovemap, speechmap, gender transposition, gender cross-coding, contrectation, feminoid, masculinoid, uxorioid, viriloid, phylism, haptoerotic, morphoerotic, gnomoerotic, paleodigm, sexosophy, homosexosophy, homosexology,* and *orgasmology,* to name a few. Clearly, Money's early identification of important concepts advanced the neonatal field of sexology. However, his passion for highly technical, esoteric terms has simultaneously burdened the field with needless conceptual deadweight and an unnecessary elitism.

Money (1988) suggests that timing and variability of sex hormones contributes to varied sexual attractions. Prenatal hormone effects of his model are listed in Table 6. Specifically, Money hypothesizes that atypical fetal hormone exposure creates differences in brain organization, and consequently, differences in gender identity, sex-typed behavior, and sexual orientation (Money, 1970, 1988; Money & Ehrhardt, 1972). Androgens are assumed to "masculinize" and "defeminize" in each dimension, while estrogens are thought to "feminize" and "demasculinize." In his model, these gendered effects can be *continuous* or *episodic and total, partial unlimited,* or *partial limited.* Previously, we noted problems inherent in linking gender roles to hormone effects. In brief, when androgens are assumed to make men "masculine" and particular social or sexual behaviors by males is defined as unmasculine, then investigators invariably look to insufficiencies in androgen exposure as the cause of the problem. The same misleading conclusion is made for women who exhibit social or sexual behaviors that are defined as unfeminine; the cause is assumed to be sex hormones. However, there is no

TABLE 5
Comparison of Two Interactionist Models of Sexual Attraction

	Kauth	**Money**
Ultimate cause	*Varied sexual attractions enhance social relationships; other-sex attraction aids reproduction, and same-sex attraction facilitates intrasex cooperation. Differential mating strategies and intrasex conflict account for sex differences in eroticism.*	*Implied but not specific. The theory cites no advantage to same-sex attraction, which is viewed as a hormonal abnormality.*
Proximate cause *Sex hormones*	*Prenatal hormones vary by timing, duration, and ratio and set erotic predispositions. Most people are in the average hormonal range and have the potential for attraction to both sexes. Above or below average exposure leads to exclusive dispositions.*	*Prenatal hormones vary by timing, duration, and ratio and set erotic tendencies. Normal hormone exposure leads to "heterosexual" attraction, and abnormal cross-sexed exposure leads to "homosexual" attraction. Attraction to both sexes is unlikely.*
Neural mechanisms	*Erotic predispositions in social information processing are shaped by cultural values, social context, and individual experience. Hypothalamus sets sexual drive; network of loci regulate sexual attraction. Reinforced erotic associations form stable erotic templates and eventually sexual orientation.*	*Hypothalamus is implicated. Appropriate hormones lead to normal behavior and "heterosexual" eroticism. Cross-sexed hormones lead to cross-gendered behavior and "homosexual" eroticism. Experience creates cognitive lovemaps of erotic imagery and ideation.*
Social experiences	*Culture influences values and behavioral opportunity, and erotic predispositions affect perception and choices through a dynamic interaction.*	*Accidental genital stimulation by same-sex parent or same-sex erotic peer play is reinforced. Cross-gendered behaviors lead to cross-gendered eroticism.*
Variability in adult sexual attraction	*Repeated reinforcement makes variability less likely over time but possible. Individuals disposed to exclusive attraction are least variable.*	*Variability is not likely.*

TABLE 6
Money's Table of Prenatal Sex Hormone Effects

	Masculinization	Demasculinization
Feminization	Androgynous	Feminine
Defeminization	Masculine	Epicene

Money's Table of Gender Cross-Coding/ Gender Transpositions

	Continuous	Episodic
Total	Transsexualism	Fetishistic transvestism
Partial unlimited	Gynemimesis/andromimesis	Nonfetishistic transvestism
Partial limited	Same-sex attraction	Bisexual attraction

published evidence that hormonal imbalances account for same-sex eroticism, paraphilias, or cross-gendered behavior.

Money (1988) referred to the effects of prenatal hormones on development as gender coding, by which he means the persistence, expression, and concordance of sex-typed traits such as gender identity, gender role/behavior, and direction of sexual attraction. His model assumes that sex-typical identification, sex-typed behavior, and "heterosexual" attraction are normative and automatic effects of the same process. Thus, when gender identity, gender role/behavior, and direction of sexual attraction differ from the norm, the individual is said to be *gender cross-coded* or *gender transposed* for one or more traits, which implies pathology. According to Money's model, boys who prefer dolls to rough-and-tumble play are cross-coded for gender role/behavior. "Gay" men and "lesbians" are cross-coded for sexual orientation, and "heterosexual" men who wear female clothing for sexual arousal are cross-coded for gender role/behavior. Money also compares the development of sexual orientation to language development; however, in his model, which language develops seems predestined. Money claims that a prehomosexual orientation is reinforced by accidental genital stimulation by the same sex parent and by prohibitions against childhood other-sex erotic play. The extraordinary series of reinforcing accidents that are required in Money's model to support homosexuality is simply improbable given the extreme homonegativism prevalent in contemporary Western culture. Yet Money claims that gender-coded predispositions are supported or modified by postnatal experiences, and once established, gender-coded traits are as immutable as native language.

In brief, the idea that variable hormone exposure creates variability in sexual attraction has merit. However, Money's model of gender-coded traits oversimplifies hormone effects, imposes a contemporary Western concept of gender, and views atypical sexual attraction and behavior as pathological. In Money's model, nonheterosexuals and people who do not engage in stereotypical sex-typed behavior are cross-coded and problematic. What is more, his table of gender transpositions is a confusing tossed-salad of gender terms that offers little clarity. Finally, Money focuses mainly on homosexuality and fails to explain how "heterosexual" attraction is predisposed or to what degree it is fixed or flexible. He implies that "heterosexual" attraction just happens.

The new model of sexual attraction views hormone exposure and sexual attraction on continuums, which allow for considerable variability. The effects of sex hormones on the development of neural systems are viewed broadly. Although some hormone effects are sex-typed, not all are. Gender is not a hormone effect; sex is. Gender, however, is defined by culture. In addition, the new model asserts that same-sex eroticism is not a pathology; rather, same-sex eroticism is a selected trait that facilitates the reduction of intrasex conflict. In sum, the new model of sexual attraction is more precise, logically coherent, and comprehensive than Money's model.

5.2.2 Variability in Sex

Nature rewards variability and diversity, and in this chapter I have attempted to demonstrate that for sexual attraction variability is inherent and evolutionarily favored. Some readers may easily accept the idea that sexual attraction is a proportion of same-sex and other-sex eroticism, whereas others may continue to resist the notion that sexual attraction is not a dichotomous trait. It is to this later group that the next few pages are devoted.

Sex is a "hard-wired" trait in the sense that sex differentiation is biologically determined and physical differences between the sexes are clear and observable. There are only two sexes, after all. Sex, therefore, can be thought of as a dichotomous trait. On the other hand, sexual orientation is a "soft" trait. Differences between people of various sexual orientations are not physical or obvious, even though our cultural beliefs tell us that people are *either* "heterosexual" *or* "homosexual." If a so-called hard-wired trait like sex turns out to be more of a continuous variable than a dichotomous one, then it becomes even more difficult rationally to dichotomize sexual orientation.

Examining the limits of sex is quite appropriate in this context; sex differentiation and sexual orientation are often conflated because they are thought to derive from the same hormonal process. By better understanding the typical variability inherent in sex differentiation and the social boundaries imposed on sex, the reader may view biological sex as less fixed and invariant.   Unfortunately, biomedical sex researchers have given little attention to typical variability in sex hormones or typical variability in sex. Researchers have been much more interested in *atypical* hormonalization and development. Fortunately, examples of variability in sex are within our reach.

We are accustomed to thinking that people are *either* male *or* female. It is not possible for an individual to have both fully formed male *and* female sex organs. Very rare individuals may have a combination of both internal reproductive tissues in the form of nonfunctional *ovotestes*, glands that have only partially differentiated (Dreger, 1998). However, most of us generally have contact with people of unambiguous sex. We easily fit people into one of two mutually exclusive categories.

Few of us commonly mistake men for women and vice versa, although we all make such mistakes occasionally. Yet despite our low error rate, there is considerable variability in the physical appearance of hormonal males and hormonal females; we are just practiced in overlooking these variations. Some men are big, bulky, and muscular; others are small and soft and feminine-looking. Some men are hairy, and others are smooth. Even so, rarely do we mistake these different looking men for women. On even closer examination, no two penises are alike. Some are big, some are small; other penises are long, and some are stumpy. Nor are testicles identical. Some men even develop breasts, and a few men lactate. By the same token, some women are slender and petite with delicate features, whereas others are large and hulking, overshadowing some men. Some women are smooth, soft, and shapely; others have mustaches and coarse body hair. Women's breasts are not identical. Some are large; some are small; some are pert, and others droop. Some nipples are large, or dark, or oval, or flat, or pointed. Breasts even vary within the pair. Vaginal and clitoral shape and length also vary from woman to woman. Yet even when women are hirsute, small-chested, and have pattern baldness, we generally have no trouble deciding that they are female.

Intrasex variance in physical morphology and appearance is common and probably due to genetic factors and typical fluctuations in sex hormones, as well as postnatal events. Indeed, maleness and femaleness each represent a *range* of physical sex-typed characteristics. Of course, social characteristics associated with gender help us de-emphasize physical variations in sex and quickly categorize people by sex. Cultural definitions of gender lead most of us to adopt, more or less, sex-typed clothing, hairstyles, speech patterns, mannerisms, and activities. Gender clues help us easily categorize people by sex, despite wide variability in physical appearances.

It is when we are faced with cases of doubtful sex that the arbitrariness of the concepts *male* and *female* and our assumptions about sex differentiation are exposed. Consider this: When is a male not *male*? What are the criteria for maleness? At what point is a female no longer *female*? What exactly constitutes *female*? There are not unambiguous answers to these questions. In real life, when the demarcation between sexes is difficult to determine, sex is usually decided by authoritative consensus (Dreger, 1998). Some examples of indeterminate sex were described in chapter 4. In rare cases, a chromosomal (XY) male is not responsive to androgen because of androgen insensitivity (AI) syndrome, sometimes referred to as testicular feminization syndrome. However, AI is a matter of degree. Not all androgen is blocked; AI is usually partial and not complete. Therefore, AI-males have a combination of male-female sex characteristics that include undescended testes, a divided scrotum, or a small penis that resembles a clitoris. Just how phallic a penis must be in order to be a penis is uncertain and open to question.   In some cases, the difference between penis and clitoris is uncertain.

In other cases of indeterminate sex, a genetic female (XX) with congenital adrenal hyperplasia (CAH) produces an abundance of androgen through the fetal adrenal glands. As a result, her body has a male-like appearance. What is more, the labia may be fused, and the clitoris is unusually large, resembling a penis. In extreme cases of indeterminate sex, sex is assigned by medical authorities and supported with hormone treatments or surgery to alter the appearance of ambiguous genitalia. However, surgical reassignment and hormone therapy have various degrees of success. Most often, if surgical reassignment is attempted, the assigned sex is female, irrespective of the degree of genital ambiguity. Making a female is easier than making a male because of the difficulty in constructing a functional, aesthetic penis.

A few cases of doubtful sex are not so evident at birth and may go unrecognized for years. For example, genetic males with 5-alpha-reductase deficiency (5-aR) lack an enzyme that is critical during fetal development in converting testosterone to dihydrotestosterone. Because testosterone cannot be fully utilized, 5-aR males look like females and have undescended testes. By most appearances, these boys are girls, until puberty. However, at puberty, the undescended testes of 5-aR males produce testosterone, and the body takes on a male-typical appearance. Because the 5-alpha enzyme is not required at puberty to process testosterone, the hormone dramatically transforms these girls to men.   Following their transformation, 5-aR individuals are reassigned sex. Even so, 5-aR men are less male-looking than men without 5-aR.

The prevalence of indeterminate sex is difficult to know. Definitions of doubtful sex are relative to the degree that genital ambiguity is unacceptable in the culture and whether *endogenous* (genetic) or *exogenous* (environmental) causes are included. In her book on hermaphroditism, Alice Dreger (1998) estimates that for every 2,000 births in the United States about one to three infants are born with "unusual anatomies" that "result in confusion and disagreement about whether they should be considered female or male or something else" (p. 42). Fausto-Sterling, a constructionist writer on gender, estimates that the prevalence of doubtful sex is even higher. She places the prevalence of doubtful sex somewhere between one in 100 and one in 1000 live births (cited in Dreger, 1998). To put these figures in perspective, Dreger notes that the frequency of doubtful sex is "about as common" as cystic fibrosis (close to one in 2,000 Caucasian births) and Down's syndrome (about one in 800 live births). With about 4 million births each year in the United States, thousands of infants are born with doubtful sex, and tens of thousands of living individuals are of an indeterminate genital sex and have been assigned a sex. Based upon genital anatomy alone, a number of people exist between the conceptual poles of *male* and *female*. These data illustrate that sex is not absolute. Contrary to popular belief, *sex is more of a continuous trait than a dichotomous one*. This is certainly not a new idea. As noted earlier, some North American Indian societies, African communities, and Pacific Islanders have recognized three genders. The three genders are male, female, and an all encompassing third-gender, composed of people with a primary same-sex erotic interest, transgendered persons, and individuals of doubtful sex (Bleys, 1995; Greenberg, 1988; W. Williams, 1986).[4]

Although cases of doubtful sex help illustrate that sex is a continuous variable, we are more likely in our every day experience to encounter individuals with fully formed, unambiguous genitalia who simply vary in masculine or feminine appearance. However, intrasex variation in physical appearance may be greater than intersex variation (Fausto-Sterling, 1992). Most of us do not meet the ideals of masculinity or femininity; we are somewhere in-between, having some "masculine" and some "feminine" characteristics. Most of us only *approximate* stereotypes of sex characteristics. For these reasons, *masculinity* and *femininity* are best viewed as relative terms and not absolute traits. Furthermore, use of gender terms like *masculine* and *feminine* is misleading, especially when applied to brain sites, general hormone effects, and sexual orientation. Although Money (1988) adds dimension and sophistication to his model by "defeminization" and "demasculinization," these concepts are still tied to a gender role model, and the same problems remain.

Common intrasex variation in physical characteristics of males and females suggests that sex hormones vary among same-sex individuals.

Although other factors may also contribute to intrasex variation in physical appearance, the argument remains that sex is not a rigidly determined or uniform trait. However, a significant degree of variance in intrasex physical characteristics of males and females can be attributed to variability in sex hormones. Hormone levels vary around a typical range, and this range varies across individuals. Unfortunately, although several studies have investigated the effects of atypical hormone exposure on fetal development, an extensive review of the literature revealed no studies that have examined typical fetal hormone exposure. Relatively little is known about the effect of typical prenatal hormone variability on development, which seems to be a field ripe for investigation.

To summarize this section, variability in physical sex-typed characteristics between, but especially within-sex supports the idea that sex is a continuous trait, even though sex differentiation is biologically determined. Wide intrasex variability in physical characteristics among males and females suggests that prenatal and postnatal hormone levels vary around a typical range and vary across individuals. This observation is consistent with the idea that variability in hormone exposure creates variability in sexual attractions.

5.2.3 Social Forces Shape Sexual Attractions

Two major forces shape human lives—the genetic stuff that we receive from our parents and the social world in which we live. These forces are independent but interactive. Together biological and cultural forces create a bubbling dynamic stew of influences where one ingredient blends into others, adding its own flavor and enhancing the mix. To understand human behavior, one must deal with both of these forces individually and where they interface—the human body.

Human infants arrive in an established social world as a set of biological inclinations and potentialities. Given a genetic preset range of trait expression, their inclinations and capacities are exploited, prized, challenged, stigmatized, or overlooked, depending upon the particular values of the culture and individual opportunities. Thus, trait expression varies within a set range, and trait expression depends largely on the environment. Like a forge, the social world pressures and molds raw but not unformed biological material into cognitions, feelings, and social behavior.

In shaping human lives, the social world provides order and meaning to experiences, influencing the kinds of encounters that individuals have and how they think about them. Social reality is created through social interactions and discourse—a reality in which people are thrust, which shapes them, and which they are able to shape to an extent but not escape. In other words, people are products of their particular social realities (Berger & Luckmann, 1966). An individual's position in the social reality, relative to others, provides him or her with a particular point of view (Foucault, 1978/1990; Padgug, 1979/1990). *As the individual personalizes and internalizes their social position and viewpoint, these become an identity.* Social identities are an adoption of social roles.

Similarly, infants are born into a social world with a set of erotic possibilities and predispositions. From the moment that they experience the world, their feelings, perceptions, and choices are influenced, and eroticism begins to take shape. In this way, sexual attraction is a dynamic product of biological predispositions, social interactions, and social definitions. Infants are not born with a sexual orientation; they develop a sexual orientation in the process of experiencing life. How people interpret their experience, given a particular social reality, defines them.

In the West, sexual feelings and personal identity are very important. We are engaged in this discussion because this era believes that understanding sexual orientation is critical to understanding the self and the behavior of others.

5.3 SUMMARY AND CONCLUSIONS

The ideas presented here are not new; the *conceptualization* is new. Other theorists have proposed evolutionary theories, interactionist theories, polygenetic theories, neurohormonal mechanisms of attraction, and the adaptive value of same-sex eroticism. However, none of these theorists have described a *general* theory of human sexual attraction that is as comprehensive. What is more, the new interactionist biopsychosocial model of human sexual attraction has several advantages over other models. The new model is more complete, logically and conceptually coherent, and consistent with historical record, cross-cultural observations, and contemporary sexual lifestyles and sexual behavior.

The new model of sexual attraction is more comprehensive that previous theories, which describe only partial, time-limited, or linear interactions between biology and the social world. Money's model (1988), for example, claims that biological processes are vulnerable to hormonal influence during a discrete critical period; thereafter, erotic predispositions are either reinforced or not reinforced by social experiences. Money claims that sexual preferences—paraphilias and such—are acquired through repeated, reinforcing behavioral experience. In his view, *interaction* seems to be unidirectional and limited. However, the interaction between biology and the social world on human experience and development is neither unidirectional nor time-limited. Biological predispositions influence behavior in a socially constructed world that limits or supports certain behavioral expressions, as well as abstract conceptions of experience. Perceptions and behaviors that are supported by the culture are likely to recur, given continued reinforcement and behavioral opportunity. What is more, biology changes with experience. When reinforced, neural pathways strengthen as a result of changes in their physical structure. Neural pathways grow and interconnect or atrophy, depending on whether they are stimulated. Thus, social and cognitive experiences, including sexual ones, produce physical changes in the brain. Furthermore, the brain is not a blank tablet, waiting to be etched; the brain arrives at birth prepared to make particular use of particular stimuli, and the significance of subsequent events leads to subsequent changes in brain structure and functioning. This is a different kind of interaction than previous theorists have described; in the new model of sexual attraction, the interaction between biology and the social world is on-going and not partial or time-limited.

Unlike past theories, the new model of human sexual attraction addresses both ultimate and proximate causation. The new model proposes that varied sexual attractions are adaptive. Consistent with evolutionary theory, variability in eroticism permits maximal adaptive responsiveness in unstable natural and social environments. Unlike previous theories of attraction, the new model asserts that same-sex eroticism plays a prominent and critical role in facilitating close emotionally significant cooperative nonkin relationships. These same-sex erotic alliances help reduce intrasex hostilities and competition, enhance social position, and acquire high-quality material, nutritional, and social resources, including access to quality sexual partners. Erotic relationships are more than cooperative friendships; erotic relationships are committed alliances bonded by emotional intimacy and they may or may not include sexual behavior. A same-sex erotic capacity is likely to have helped hominid females survive and successfully raise children to adulthood in primarily all-female communities of nonkin. Women needed close alliances with other females to ensure social position, defend against hostile females, access quality food, and assist with child rearing. Those females who successfully navigated these dangers passed on their genes and their traits to successive generations. Life for early hominids, and indeed for modern humans until recently, was sex-segregated. Hominid males needed loyal alliances to defend against male aggression, coordinate fighting and hunting, and gain access to quality female sexual partners. Single males without allies were disadvantaged; cooperation—loyal cooperation—ensured social and reproductive success. Those males who survived these natural and social dangers passed on their genes and their traits to their progeny.

Contemporary men and women are benefactors of their ancestors' success. Thus, it is likely that most people have the erotic capacity for emotionally significant intimate relationships with someone of the same-sex, whether or not they exercise it. Culture and the strength of the erotic trait determine to what extent eroticism is sexual. That sexual behavior is a product of same-sex erotic relationships is neither improbable nor necessary. On the other hand, those few individuals who have an exclusive same-sex eroticism have less flexibility in sexual feelings than do those with an undifferentiated eroticism.

Given that sexual attraction is a polygenetic heterozygous trait, the new model of attraction proposes that exclusive other-sex and same-sex eroticism are functionless by-products of typical genetic variation. Although each form of eroticism has limited function, neither form enhances reproductive success over undifferentiated eroticism. In this model, varied erotic attractions optimize reproductive success. *Function* here is used in the evolutionist sense of adaptation and reproductive success; *function* does *not* refer to a social value. Value for evolutionists means adapting in such a way that reproduction is most successful. Social values are defined by culture, not biology. Exclusive other-sex or same-sex attraction could have great or little social value, depending on the culture. This is certainly the case in contemporary Western society; exclusive other-sex attraction or heterosexuality is socially privileged. The social advantages or disadvantages attributed to exclusive sexual attraction is culturally dependent and can vary widely. Exclusive sexual attractions have *relative*—not inherent—social value. This distinction deserves emphasis because the new model of sexual attraction is one of few theories that does not hold exclusive other-sex (male-female) erotic attraction in a privileged biological *and* social position.

As noted earlier, the new model of human sexual attraction proposes that attraction traits are polygenetic and heterozgyous. Genes associated with sexual attraction contribute to variability in the proportion, timing, and duration of prenatal sex hormone exposure, as well as in postnatal levels of circulating hormones. Neurohormones are the mechanism for trait expression. For most people, hormone exposure occurs within a typical range, although the pattern varies. Variability in sex hormones is one factor that accounts for significant variance in intrasex physical differences. Although the same hormonal mechanisms produce sex differentiation and erotic predispositions, these phenomena are separate and distinct. Sex differentiation and the establishment of erotic predispositions have different critical periods. Being *male* or being *female* is not synonymous with other-sex eroticism, and so-called cross-sexed hormone exposure (Money, 1988) has not been found to account for exclusive same-sex eroticism. Because sex differentiation occurs on different timetables for males and females, it is plausible that erotic predispositions are also set during different critical periods. Compared with development of external genitalia, which takes about six weeks, brain development is a relatively slow process, extending beyond birth and allowing for a wide range of hormonal and experiential influences for a considerable period of time. Thus, the potential for variability in brain-linked traits is greater than in physical or genital appearance. Consistent with this observation, sexually dimorphic nuclei in the brain show considerable intrasex variance (LeVay, 1991, 1993), and sex-typed patterns of cognitive functioning such as lateralization and hemispheric dominance also have great intrasex variability, regardless of sexual orientation (Gladue *et al.*, 1990; McCormick & Witelson, 1991).

Consequently, the new model of sexual attraction refutes the use of absolute gender terms like *masculine* or *feminine* to describe hormone effects on the brain. These terms are culturally biased and conceptually limited. Sex hormones and gender roles are not synonymous. Too often, the terms *masculine* and *feminine* have been used to suggest that a "heterosexual" gender role results automatically and inflexibly from exposure to *normal* androgens and estrogens. Theorists who employ gendered models of sexual attraction such as Money (1988) find themselves describing layers upon layers of appropriately sexed or cross-sexed effects and traits, which adds no specificity to the model and, worse, creates differences where none exist. This, in fact, is the trap of gendered models of attraction—that all deviations from the normative standard are explained as cross-sexed effects. Theorists who use gendered models of attraction seem unaware that *gender* and *gender role* are culturally dependent.

The new model of sexual attraction purports that patterns of prenatal hormones create erotic predispositions in social information processing systems. Variability in neural sensitivities to competing sex hormones allow for overlapping critical periods and a complex tapestry of erotic predispositions and sex-typed and other traits. Contrary to previous theories, the new model holds that no single brain site governs sexual attraction. The hypothalamus is more likely to regulate sex drive or motivation than sexual attraction. A network of brain structures involved in sexual attraction includes the amygdaloid bodies, hippocampus, association cortex and prefrontal cortex. Researchers who report differences in single brain sites between people of different sexual orientations exaggerate their significance. Isolating differences in one element of a larger network reveals little about how the system works.

Similar to Money's (1988) concept of lovemaps, the new model of sexual attraction proposes that erotic predispositions lead to reinforced cognitive and social experiences that eventually form stable erotic templates—precursors of sexual orientation. Over time, erotic templates, as practiced habits, become resistant to alteration. However, unlike Money's lovemaps, the new model asserts that sexual orientation is resistant but not invulnerable to alteration. Given behavioral opportunity and significant social circumstances, most people are capable of a variety of erotic experiences. Even so, unless adults have denied themselves sexual expression, by middle age variation in sexual orientation is infrequent.

Unlike social constructionist theories, the new scientific model of sexual attraction describes human beings as situated in a rich abstract social context, inseparable from their physical bodies, and influenced by their biology and genetic history through an active, dynamic interaction. The human body is the point in which biology and social experience interface and interact, and as such, the body and this interaction cannot be left out of a general theory of sexual attraction. Indeed, human sexuality is the product of this interaction.

For the reasons presented here, the new interactionist biopsychosocial model of sexual attraction represents an improvement over current models. As a scientific theory, this one will prompt discussion, generate alternative hypotheses, stimulate research, and find support or not. A practical parsimonious theory must undergo the scientific process of critical peer review and confirmation. Whether or not this model of sexual attraction is ultimately confirmed and accepted, my goal has been to provide a new conceptualization of sexual orientation that contributes to a fuller understanding of human behavior. On this count, I believe that I have already succeeded.

However, before drawing this text to a close, I will propose a research program that may substantiate or refute this model of sexual attraction. Such a research program is the topic of the final chapter.

CHAPTER 6: PRESENT AND FUTURE QUESTIONS

A good theory not only summarizes and incorporates extant knowledge but is heuristic in that it originates and develops new observations and new methods. — Theodore Millon, *Normality: What may we learn from evolutionary theory?*

Previous chapters described a new interactionist model of sexual attraction in which erotic predispositions in social information processing guide, shape, and are shaped by postnatal social experiences. The differences between the new model and current models were examined. I argued that the new model of sexual attraction better fits current data and cultural observations than do other models.

Although the new model of sexual attraction appears to have conceptual advantages over others, a critical test of the model's utility is whether it holds up under empirical scrutiny and generates productive research. A good model should not only answer research questions; it should also raise them. In this final chapter a limited research program for testing the new model of sexual attraction is outlined.

6.1 A RESEARCH PROGRAM

Interactionist theories are complicated and difficult to test. Even so, my wish is that this new model of attraction will stimulate critical thought about sexuality and promote good empirical research. Although scientists generally acknowledge that human sexuality is multi-determined, most sexuality research study designs assume a single cause. The challenge for researchers is to test an interactionist model of sexual attraction.

To test the new model of sexual attraction, initial research efforts should be directed toward three particular goals. One of these goals is the creation of *a measure of eroticism*. Although eroticism can include sexual behavior, it is a more encompassing concept. A measure of eroticism might assess to what degree each sex is a central figure in actual and ideal or preferred sexual desire, sexual fantasy, romantic fantasy, emotional intimacy including close affectionate nonkin relationships and mentor-type relationships, and social affiliation. Patterns of responses to this assessment might likely identify various erotic subtypes. At any rate, a good measure of eroticism—separate from sexual behavior—is essential in validating the new model of attraction.

A second goal of initial research efforts should be the development of *a measure of sexual attraction*. Because sexual attraction is a component of general eroticism, a sexual attraction factor could be included in a global eroticism measure. A sexual attraction factor might assess the degree to which each sex is a central figure in several actual and preferred domains. These domains could be sexual fantasy, romantic fantasy, a desire for physical closeness and emotional intimacy, romantic love, lust, committed partnership or ideal mate, special affectionate relationships, and sexual intimacy through casual sex or regular sexual relationships. Sexual behavior is included last in this list in order to de-emphasize its importance. Sexual behavior is not always consistent with sexual attraction and therefore should be one of several measures of attraction. Perhaps more than other domains, explicit sexual behavior is subject to social regulation and behavioral opportunity. Still it is worth noting in future studies whether sexual behavior is consistent or inconsistent with other dimensions of sexual attraction.

Sexual orientation was deliberately omitted from the assessment of sexual attraction. For reasons already discussed, self-identity is also not a good indicator of sexual attraction. In brief, categories of sexual orientation represent a particular Western viewpoint and model, which may have no applicability to non-Western cultures. Sexual identity categories, like sexual behavior, are strongly influenced by social and political forces and consequently may not be consistent with sexual attraction or sexual behavior. According to the new model of attraction, how one defines oneself is not necessarily reflective of sexual attraction, although given social values and concepts shape the experience of attraction.

The two measures briefly described here are intended as guides for instrument development, not blueprints. Both measures portray general eroticism and sexual attraction as multi-dimensional. Thus, results of these measures will yield an erotic or sexual attraction *pattern* or *constellation* rather than provide a single score or identify a discrete category of person.

A third critical goal of initial research efforts in the investigation of the new model of sexual attraction is the promotion of *cross-cultural studies*.

Except for anthropological observations, most empirical research in sexuality has been conducted on samples of Western men and women. Contemporary Western attitudes and behaviors do not represent or even reflect the whole of human experience. When the only research data available is from Western samples, theorists inadvertently create Western-biased models of sexual attraction that have little or no relevance to non-Western cultures. Anthropological observations from primarily Western observers suffer a similar problem. This is not to say that Western anthropologists and researchers have a political or social agenda and that non-Western scientists somehow have removed such biases from their observations. As discussed earlier, despite rigorous training and efforts at objectivity, scientists are still products of their culture and as such reflect particular assumptions and cultural points of view. Although it may not be possible to eliminate cultural assumptions from scientific inquiry, including a variety of native and non-native observations and balancing observations from observers of different cultural backgrounds may help to minimize single cultural viewpoints and reveal a truer picture of universal human sexuality. For this reason, *many current, well-known studies of sexuality need to be validated on non-Western samples in order to have general relevance.* Biological studies that find support among non-Western peoples enhance the significance of their findings. Almost every area of sexuality research discussed in this text could benefit from cross-cultural validation.

Attention to these initial goals would advance not only the investigation of the new model of sexual attraction but also advance sexuality research in general. Once the above tasks have been addressed, the next step in a test of the new model is to examine its predictions. In the sections below, predictions from the new model of attraction are applied to four general areas of sexuality research—genetics, hormones, cognitive functioning and information processing, and developmental studies. Each is discussed in turn.

6.1.1 Genetics Studies

The new model of sexual attraction predicts that genetic variation produces three basic erotic types. (A rare fourth type represents no sex-specific sexual attraction). The three erotic types are *exclusive other-sex eroticism, exclusive same-sex eroticism,* and *undifferentiated eroticism.* As discussed in chapter 5, multiple genes are thought to direction eroticism, and undifferentiated eroticism reflects a heterozygous genotype. Exclusive eroticism represents homozygous forms of the trait. Thus, patterns of dominant *and* recessive forms of the trait contribute to undifferentiated erotic types, and patterns of dominant *or* recessive forms of the trait contribute to exclusive erotic types.   Furthermore, exclusive (homozygous) erotic genotypes *should* correspond with exclusive sexual attraction in adulthood. Undifferentiated (heterozygous) erotic genotypes *should* more often correlate with the socially accepted form of sexual attraction; however, undifferentiated erotic types *should* show greater variation in erotic relationships than exclusive erotic types.

This line of research supposes that it is possible to identify specific genes associated with eroticism.   At present, however, this is not the case. Hamer and associates (1993, 1994) have come the closest to date to identifying genes associated with eroticism.   Hamer identified five genetic markers for maternally transmitted same-sex eroticism in males.   Even so, this type of work is in its infancy, and Hamer's findings need to be confirmed by independent researchers.   The five markers encompass hundreds of genes.   In addition, other genetic forms of same-sex eroticism may exist, and genes associated with same-sex eroticism could reside in places other than at Xq28. It is also likely that the location of genes associated with eroticism differ for males and females (Pattatucci & Hamer, 1995).

Genetic researchers who investigate sexual attraction may choose to begin by first searching for exclusive (homozygous) erotic types among each sex. Same sexed individuals with a similar exclusive sexual attraction *should* have similar versions of genes in corresponding loci. For example, men with an exclusive same-sex erotic attraction should each have the same recessive versions of genes, and on the average, these men should show a high correlation for recessive genes in the same locations. The same should be true for women with an exclusive same-sex erotic attraction, except that the genes in question are likely to be in different locations than for men.   Because exclusive same-sex eroticism is infrequent and because of stigma associated with homosexuality, people who acknowledge experiencing same-sex eroticism are probably less likely to misrepresent their erotic experience; therefore, researchers may have an easier time selecting an appropriate sample and identifying genes associated with this trait.   Genes associated with exclusive other-sex eroticism—because only the dominant versions are represented—*should not* correspond the genes related to exclusive same-sex eroticism.   However, individuals with an undifferentiated eroticism should show a mixed genotype pattern relative to homozygous genotypes.

After erotic genotypes have been identified, genotype studies must be validated with cross-cultural populations. Similar results among non-Western samples would provide substantial support for the biology of human sexual attraction—specifically, that genes direct erotic predispositions.

Although Science lacks the technical skill at present to identify specific erotic genotypes, the probability that this knowledge will exist in the near future raises several serious ethical concerns.   Edward Stein (1999) provides an excellent discussion of these concerns. He notes that individuals who have an exclusive same-sex erotic attraction are simply less valued in Western culture than are people who have an other-sex erotic attraction. In our present heterosexist society, information about an unborn child's potential erotic genotype could lead parents to abort a fetus that may grow up to be "gay" or "lesbian." (This situation presents an interesting dilemma for anti-abortionists, many of whom also hold conservative religious beliefs that condemn homosexuality.)[1] Is abortion of a potentially "gay" fetus a lesser evil than abortion of a potentially "heterosexual" fetus?   Of course, knowledge of erotic genotypes suggests more certainty about trait expression and behavior than is possible and trivializes the influence of social experiences on trait expression. However, Stein convincingly argues that—given this culture's strong homonegative-heterosexist values—even an ineffective genetic test that purports to identify "gay" fetuses would be popular and financially successful. According to Stein, the "beauty" of the genetic test-scheme is that it plays upon the fears of parents and the test's ineffectiveness would be nearly impossible to prove. Aborted fetuses cannot grow up to demonstrate any kind of sexual attraction, and parents may be unwilling to take even a small chance that their child could be "gay." Some parents might reason that the mistaken abortion of a few "heterosexual" fetuses is an acceptable risk in order to avoid having "gay" kids.

On the other hand, if a supposedly "gay" fetus is not aborted, knowledge about the child's possible adult sexual attraction is likely to affect not only parental decisions about how the child is raised but also the child's own self-perception. A genetically identified "gay" child is likely to be treated differently than supposedly "heterosexual" children, and not for the better. A genetically identified "gay" child who grows up to be "heterosexual" may always be under a cloud of suspicion. Proponents of a genetic test could also claim that such an individual received a rare false positive test result.

As long as homosexuality is stigmatized, research on erotic genotypes presents an enormous ethical and moral precipice. Unfortunately, as is usually the case, rapid advances in biomedical technology will confront society with these dilemmas well before we fully appreciate the consequences of such knowledge and our subsequent actions.

6.1.2 Hormone Studies

The new model of sexual attraction predicts that prenatal sex hormones vary in ratio, timing, and duration within a typical range.

Verification of this assumption is a critical step in testing the new model. Patterns of fetal hormone exposure are hypothesized to predispose erotic information processing and contribute to adult sexual attraction. Very little is known about typical variability in prenatal sex hormones or their effect on development. Little is known, as well, about the effect of estrogen on typical development. Long-term prospective studies should assess these hormone effects from the womb to at least young adulthood, particularly as they relate to eroticism and sexual attraction.

In addition, the new model of sexual attraction predicts that sex hormones have significant effects in particular areas of the brain, including the amygdaloid bodies, hippocampus, prefrontal cortex, and association cortex. Specifically, competing sex hormones affect neural density, sensitivity, and dendritic growth in selected areas of the brain, which are thought to process erotic stimuli. Neural structures in these areas should be examined for sex differences and differences by sexual attraction. The amygdaloid bodies are thought to initially assign erotic significance to stimuli and therefore may be especially sensitive to sex hormone effects. Furthermore, studies should investigate the effect of social experience on neural pathways in these brain sites. Evidence that social experiences alter neural morphology at key sites would further support the new model of attraction. The model also predicts that selective hormone effects on the social information processing system predispose erotic associations. Brain scans may identify which neural systems are active when individuals are presented with erotic and sexual stimuli of different types. Different erotic types of individuals should not differ in the systems that process erotic stimuli, although sex differences are likely in how these systems are employed.

Although the new model of sexual attraction makes no specific predictions about the affect of sex hormones on the inner ear, recent findings about differences in otoacoustic emissions between people of different sexual orientations should be replicated with cross-cultural samples. Consistent findings across cultures would further reduce the likelihood that variations in otoacoustic emissions between people of different sexual orientations are related to social experience.

Positive findings from studies related to pheromones and immune system detection support the contention that sexuality has a biological foundation. Hormonal communication may reflect an ancient mating strategy or a form of mate selection. Sexually active adult men benefit by detecting fertile women, and women reduce the differential reproductive advantage of proximal women by synchronizing their menstrual cycles. However, current findings do not state whether biological communication is independent of sexual attraction. That is, are "lesbians" likely to synchronize their menstrual cycles with neighboring women, and do "lesbians" respond favorably to female pheromones? Similarly, do "heterosexual" and "gay" men respond differently to male pheromones? Positive findings in this area would provide supportive evidence that sexual attraction communication is biology and that people respond to sex-specific biological messages, consistent with their sexual attraction. Again, cross-cultural studies are essential in order to generalize such findings.

In addition, evidence of hormonal communication could provide supportive evidence for the significance of same-sex erotic alliances. If same-sex erotic alliances are important as the new model of attraction predicts, special intimate friends may also communicate hormonally much like sexual mates. For example, emotionally significant erotic partners may have dissimilar immune systems similar to preferred sexual mates.

6.1.3 Cognitive Functioning and Information Processing Studies

The new model of sexual attraction predicts that erotic predispositions are evident in prepubertal children and adolescents. Cognitive studies should investigate the presence of erotic predispositions among children and adolescents, as well as the role of erotic significance for each. Speed of erotic information processing and recall of preferred sex-typed erotic words could provide supportive evidence of erotic predispositions. Ideally, prospective studies would follow adolescents with stable erotic templates to adulthood. Evidence that early sex-specific predispositions correspond to adult sexual attraction would support predictions by the new model of attraction.

Several studies have found that men and women process erotic information differently. In a review of this literature, Geer and Manguno-Mire (1996) noted that men and women demonstrate an *appraisal bias* by responding more slowly to stimuli when erotic content is present. However, men are faster than women at identifying sexual stimuli and more likely to rate these stimuli as more positive. On the other hand, women appear to hesitate when exposed to sexual stimuli, perhaps to allow more time to process information. Yet women recognize romantic stimuli faster than men and rate these stimuli as more positive. In addition, Manguno-Mire and Geer (1998) found that semantic information appears to be cognitively organized in a "net." Women typically have complex organizations of relationship-oriented words, whereas men have complex networks of sexual words. Manguno-Mire and Geer noted that sexual information is processed differently than nonsexual information and that processing differs not only by sex but also by sexual orientation. Cognitive selectivity of sex-specific stimuli and content lend support to the hypothesis that predispositions in erotic information processing lead to development of varied sexual attractions.

Neuroimaging scans may also detect erotic predispositions during exposure to preferred and nonpreferred erotic stimuli. Selective neural responsiveness to sex-specific stimuli would lend support to the new model of sexual attraction. What is more, scans may find that different neural pathways process different erotic experiences and that the number and complexity of pathways differ between undifferentiated and exclusive erotic types. That is, individuals who have an undifferentiated erotic capacity may demonstrate more numerous and complex erotic processing pathways than do people who experience exclusive attractions. Because social experiences and behaviors may affect the size and functioning of neuroanatomic structures, subjects may need to be categorized by degree and type of sexual experience.

Finally, neuroimaging studies of erotic information processing may also demonstrate that sexual attraction is primarily a cortical function and not a subcortical activity.

6.1.4 Developmental Studies

The new model of sexual attraction predicts that erotic predispositions are evident among children around late childhood, but no earlier than age eight. Therefore, prior to puberty children may begin to demonstrate erotic preferences through childhood "crushes," fantasy play, and emotional attachments to nonkin adults or fictional persons. Given our current heterosexist culture and for the reasons noted earlier, researchers may have an easier time identifying an exclusive same-sex erotic predisposition among late childhood boys and girls than identifying exclusive other-sex eroticism. In the present Western culture, children and young adults who have same-sex erotic feelings often recognize at an early age that they are different from their peers (Malyon, 1981), and retrospective reports note that most "gay" men, "lesbians," and "bisexuals" experience same-sex erotic feelings during adolescence (Fox, 1995).

By middle adolescence, erotic predispositions and preferences may be evident. Adolescents with an undifferentiated eroticism may form significant same-sex erotic partnerships as well as manifest erotic attraction to the other-sex. Collateral reports from parents or other adults could help substantiate self-reports. However, prospective studies are necessary to determine that an early pattern of erotic predispositions continues into adulthood. Given the current heterosexist culture, most adults should have significant same-sex friends and other-sex sexual partners. On the whole, people who have the capacity for undifferentiated eroticism should have diverse social networks and significant erotic relationships with both sexes. On the other hand, people with an exclusive erotic attraction should more often form affectionate intimate relationships with others of the same sex. People who have an exclusive same-sex eroticism should report few emotionally significant other-sex relationships, whereas people with an exclusive other-sex eroticism should show the reverse pattern. A sex preference in social affiliation should be detected during adolescence.

The new model of sexual attraction also predicts that men and women employ same-eroticism differently. Consistent with evolutionary demands, men should use same-sex eroticism to form close male alliances to manage intrasex aggression and to facilitate the acquisition of quality resources, including sexual partners. For most men, a close male partnership should be invested with an affectionate and sentimental quality and accompanied by expectations of emotional fidelity and feelings of jealousy, which sets the relationship apart from casual friendships and acquaintances. For men, a special male partnership should serve to bolster self-confidence, strengthen defenses against threats from hostile males, facilitate material success, and contribute to higher social standing in the community. In short, men with close male alliances should be more socially successful than men without close male alliances or men with only female alliances. Said another way, men with an undifferentiated eroticism should have close, loyal male friends; they should be materially successful, socially skilled, charming, and comfortable with women relative to men who have no close male friends. It is also probable that such men get their emotional needs met through their male friends rather than with female sexual partners. Ultimately, men with an undifferentiated eroticism should produce more offspring who reach childbearing age on the average than do other men. If this hypothesis is correct, women should view men who have close male friendships as more sexually attractive and socially successful than men who lack close male relationships. In addition, women should rank men with an exclusive erotic attraction, whether for females or for males, as less sexually attractive and less preferable as a marriage partner compared with men who have an undifferentiated eroticism.

The new model of attraction predicts that women use same-sex eroticism to form emotionally close female partnerships in order to reduce intrasex hostility, provide assistance during child rearing, and obtain high-quality material and nutritional resources. Therefore, for most women a significant female partnership should be invested with an affectionate, romantic quality marked by emotional fidelity, proximity, and jealousy. Women who have close female friends should share child-care duties, have higher social standing in the community, and should be less vulnerable to hostilities from other women than women who lack close female friends. Social standing may provide women more choice in male sexual partners, when choice is possible. Similarly, women who have the capacity for undifferentiated eroticism are more likely to get their emotional needs met by women than by men. Ultimately, women with an undifferentiated eroticism should raise more offspring to childbearing age on the average than do women with exclusive sexual attractions.

Because contemporary Western life is so different from the conditions in which varied sexual attractions evolved among our hominid ancestors, anthropological observers must test these predictions in small isolated non-Western sex-segregated communities. Unfortunately, Western culture has already influenced many of these communities, making differences between cultures less distinct.

To summarize this section, this limited research program proposes several projects in genetics, hormonalization, cognitive functioning, and development that will help to validate or disconfirm the new model of sexual attraction. Initially, researchers must develop reliable multi-dimensional measures of general eroticism that include sex-specific erotic attraction. As described, these measures would portray an erotic and sexual attraction pattern rather than report a single score, label, or category of person. Lastly, future research efforts must validate study findings among non-Western cultures in order to generalize results.

6.2 AFTERWORD

In this text, readers have traversed a vast territory. We have discussed a variety of theoretical models of sexuality from Darwin to Dörner, from Freud to Foucault, and from Plato to Hamer. We have examined a variety of sexual cultures from prehuman to ancient Babylon, ancient Greece and Rome, premodern Japan, premodern Arabia, Western society in the late 19th and 20th centuries, Melanesia, Africa, and North American Indians. My hope has been that in this journey readers will see part of the larger pattern that comprises human sexuality.

Nature is diverse, and human beings share in that diversity; indeed, we are products of it. This text noted that our hominid ancestors faced difficult challenges in order to survive and adapted sexual attraction for this purpose. We discussed the diversity of human sexual cultures and the many remarkable similarities between them. We considered how cultural beliefs shape perception, influence experience, and bias observations. We critiqued disparate theories of human sexuality, revealing their underlying assumptions and testing their logic. Some theories found limited support, whereas others failed to fit the data of human experience.

What emerged from this discourse was a new model of sexual attraction, grounded in evolutionary principles and built on a dynamic interaction between biological traits, cultural context, behavioral opportunity, and individual experience. An interactionist account of these factors best explains the regular pattern of sexual attractions that are observable across cultures and throughout history. From this new perspective, human sexual attraction is seen as an active interface between biology and the social world. Theories of human sexuality that omit the physical body or disregard culture or fail to account for behavioral experience present a stilted view of human life. Theories of sexual attraction must reflect human experience and not redefine "square" experience to fit a preconceived "round" model.

Although this text is ending, the project of this book is not nearly complete. That is not a failure, however. In challenging conventional conceptions of sexuality and expanding the boundaries of sexual theories beyond homosexuality, I hope to have offered a new and promising perspective on sexuality. I hope to have stimulated creative thought and ignited curiosity. I hope to have answered a few questions and raised others.

In my mind, a good theory is a work in progress; it is never a finished product. Thus, a good theory is not an end. It is a starting place.

Endnotes

Chapter 1: Obfuscation and Clarification: An Introduction

1. Edward Stein (1999) makes similar arguments in *The Mismeasure of Desire: The Science, Theory, and Ethics of Sexual Orientation*. He devotes much of his book to a critique of what he calls the "emerging scientific program for the study of sexual orientation." By this he means recent genetic, hormonal, or neuroanatomic research on same-sex erotic attraction in which sexual orientation is viewed largely as a dichotomous trait.

2. Time and space prohibit a complete discussion of each definition of *natural*. The interested reader is referred to Boswell's 1980 book *Christianity, Social Intolerance, and Homosexuality*, much of which is devoted to explicating the various meanings of *natural* and *unnatural*.

3. Kauth and Landis (1996) provide a thorough critique of Butler's argument against granting oppressed minority status to "gay" men and "lesbians" and not allowing them to serve openly in the United States military. Butler's argument is based on the assumption that same-sex erotic behavior is acquired or learned.

4. Didi Herman (1997) provides an excellent description of the Christian Right's conceptualization of "gay" men and "lesbians" and how this concept fits with their political goals.

5. For various historical definitions of sodomy, see Irving J. Sloan (1987), *Homosexual Conduct and the Law*; John Boswell (1980), *Christianity, Social Intolerance, and Homosexuality*; and Lynn Witt, Sherry Thomas, and Eric Marcus (Eds., 1995), *Out in All Directions: The Almanac of Gay and Lesbian America*.

6. Forster wrote *Maurice* in 1913 but left the manuscript unpublished in his desk drawer, fearing negative public reaction to the novel's homoerotic content. *Maurice* was published many years later, posthumously.

Chapter 2: Libido and Learning: Psychosocial Models

1. I am grateful to David Powell for suggesting this imagery.

2. Bayer (1987) also describes the social implications of diagnosing a whole class of people as mentally ill.

3. Actually, a nude woman would also be a CS. However, a nude woman fondling a male subject's penis could be an unconditioned stimulus (UCS). In addition, erections can occur without an apparent stimulus. Men, especially adolescent males, often report spontaneous erections to nonsexual stimuli such as being punished, being chased, or witnessing a fire or

accident (Bancroft, 1970). In these cases, erection may result from general physiological arousal.

4. Many species are predisposed to make particular nonsexual associations but not others (Weinrich, 1987a). Pigeons, for example, quickly associate pecking with reward but resist learning to peck to avoid punishment. Newly hatched chicks willingly tolerate aversive stimuli in order to be near their mother. Rats are poor visual discriminators but easily learn fine olfactory discriminations.

Chapter 3: Culture's Child: A Social Constructionist Model

1. Social change alters cultural beliefs. A few years after Plato's *Symposium*, Xenophon countered Plato's idealization of male love with his own *Symposium*, which championed the love of women and marriage (Cantarella, 1992). After the death of Socrates for corrupting the youth of Athens and the defeat of the city in the Peloponesian Wars, the social climate of Athens changed dramatically. A wave of conservatism and asceticism swept the city. Plato was not unaffected by the changing political and intellectual clime and fled Athens after Socrates' death. In his last work, *Laws*, he rejected erotic pleasure as distracting and embraced asceticism. Plato wrote that only sexual pleasure directed toward procreation was consistent with nature (*kata physin*). Nonprocreative sexual activities, whether with males or females, were now viewed by Plato as counter to nature (*para physin*). It is unclear whether his later views were a final evolution of ideas or a self-protective reaction to the reality of the conservative politics of the day.

2. Herman (1997) describes how Christian Right revisionists present American history as a struggle for a theocratic Christian state. Christian Right revisionists portray the Founding Fathers as devout Christians who never intended a separation of church and state. Leading Christian Right revisionists include David Barton, president of the Wallbuilders; Peter Marshall; Gary DeMar; John Whitehead, president of the Rutherford Institute; and Rousas John Rushdoony. Christian Right revisionist "historians" are frequent guests on Pat Robertson's 700 Club, D. James Kennedy's televangel show, and James Dobson's Focus on the Family radio program.

3. I am grateful to Patrick Hopkins for suggesting this possibility.

Chapter 4: Parts and Pieces: Evolutionary and Biological Models

1. To be fair, Money (1988) hypothesizes that "masculine" and "feminine" gender transpositions vary on several dimensions—total or partial unlimited or partial limited, continuous or constant, and episodic or alternating. However, these dimensions are more confusing than clarifying. Money's gendered model is discussed in more detail in chapter 5.

2. Thanks to David Gochman for reminding me of this fact.

3. Although rare, some individuals have more or less than two sex chromosomes. Turner syndrome is characterized by having one X chromosome or three or more. Such individuals look like females. Klinefelter's syndrome is characterized by an XXY or XYY chromosomal pattern. These individuals are male in sex and appearance. The behavioral consequences of these genetic disorders are beyond the scope of this book.

4. For radical feminists who view testosterone as the cause of aggression, competition, oppression of women, and many of the world's problems, it may be disconcerting to discover that estrogen compounds "masculinize" male brains.

5. Dabbs (1990) notes that men typically experience a 55% percent drop in testosterone level from morning to night.

6. McKnight (1997) provides an excellent discussion of same-sex erotic attraction as a population-balanced polymorphism. These ideas are complex and not without controversy. Different genetic models predict different rates of same-sex attraction in the population, including elimination of the trait.

Chapter 5: A Synthesis: The Development and Function of Sexual Attractions

1. Contraceptive strategies were not unknown in the past, although they were far less effective and convenient by today's standards. In many cultures, women delay or avoid conception and space child births through prolonged nursing for one or more years.   Other common contraceptive strategies include ejaculation outside of the vagina (*coitus interruptus*), anal intercourse, postnatal bans on sexual activity between couples, intercural intercourse (between the thighs), and prohibitions against intercourse because of ritual impurity. However, some ethnic and cultural prohibitions worked against contraception by restricting intercourse to those periods of time that would most likely result in conception.
2. A related problem is the common practice among researchers of combining "bisexuals" with "gay" men and "lesbians," based on their same-sex sexual behavior. In fact, "bisexuals" may share few social or constitutional characteristics with "gay" men and "lesbians."
3. These examples are borrowed from Weinrich (1987a).
4.   In most of these cultures, third-gendered individuals engaged in cross-gendered roles, behaviors, and affections. It is unlikely that many third-gendered individuals had ambiguous genitalia.   The idea that people attracted to others of the same-sex were cross-sexed and members of a third-sex or intermediate sex enjoyed popularity in Western culture during the late 19th and early 20th centuries (Bleys, 1995).

Chapter 6: Present and Future Questions

1. Family conflict over knowledge about a fetal gay genotype was portrayed dramatically in the 1993 Broadway play and 1997 film, *The Twilight of the Golds*. Fictional genetic tests reveal that daughter Suzanne's unborn son will be gay. Suzanne's brother David is gay. The Gold family has suppressed their discomfort with his David's sexuality until the baby's test results are revealed. Several family members including the baby's father encourage Suzanne to abort the fetus. David sees the family's willingness to abort the baby as an unspoken wish that he had not been born. David's parents say as much during a heated argument that exposes the family's shame and guilt, shattered expectations, and desire to blame someone else for their disappointments.   Eventually, the Gold family—with the exception of the baby's father— decide that David and the baby are more important than their expectations about how people and life should be.

Bibliography

ACSF Investigators. (1992). AIDS and sexual behaviour in France. *Nature, 360,* 407-409.

Alexander, J. E., & Sufka, K. J. (1993). Cerebral lateralization in homosexual males: A preliminary EEG investigation. *International Journal of Psychophysiology, 15,* 269-274.

Alford, G. S., Plaud, J .J., & McNair, T .L. (1995). Sexual behavior and orientation: Learning and conditioning principles. In L. Diamant & R. D. McAnulty (Eds.), *The psychology of sexual orientation, behavior, and identity: A handbook,* (pp. 121-135). Westport, CT: Greenwood Press.

Allen, L. S., & Gorski, R. A. (1991). Sexual dimorphism of the anterior commissure and massa intermedia of the human brain. *Journal of Comparative Neurology, 312,* 97-104.

Allen, L. S., & Gorski, R. A. (1992). Sexual orientation and the size of the anterior commissure in the human brain. *Proceedings of the National Academy of Science, 89,* 7199-7202.

Allman, W. F. (1994). *The stone age present.* New York: Simon & Schuster.

American Civil Liberties Union. (1999, January). Lesbian and gay rights: State sodomy statues. http://www.aclu.org/issues/gay/sodomy.html

American Psychiatric Association. (1998, December 11). Position statement on psychiatric treatment and sexual orientation. http://www.psych.org/news_stand/rep_therapy.html

American Psychological Association. (1997, August 14). Resolution on appropriate therapeutic responses to sexual orientation. http://www.apa.org.pi/lgbpolicy/orient.html

Anderson, D. K., Rhees, R. W., & Fleming, D. E. (1985). Effects of prenatal stress on differentiation of the sexually dimorphic nucleus of the preoptic area (SDN-POA) of the rat brain. *Brain Research, 332,* 113-118.

Arendash, G. W., & Gorski, R. A. (1983). Effects of discrete lesions of the sexually dimorphic nucleus of the preoptic area or other medial preoptic regions on the sexual behavior of male rats. *Brain Research Bulletin, 10,* 147-150.

Baal, J. V. (1966). *Dema, description and analysis of Marind - anim culture.* The Hague: Martinus Nijhoff. (Original work published 1934).

Bailey, J. M., & Pillard, R. C. (1991). A genetic study of male sexual orientation. *Archives of General Psychiatry, 48,* 1089-1096.

Bailey, J. M., Willerman, L., & Parks, C. (1991). A test of the maternal stress theory of human male homosexuality. *Archives of Sexual Behavior, 20*(3), 277-293.

Bailey, J. M., & Pillard, R. C. (1993). Reply to a "Genetic Study of Male Sexual Orientation." *Archives of General Psychiatry, 50,* 240-241.

Bailey, J. M, Pillard, R. C., Neale, M. C., & Agyei, Y. (1993). Heritable factors influence sexual orientation in women. *Archives of General Psychiatry, 50,* 217-223.

Bailey, J. M., Gaulin, S., Agyei, Y., & Gladue, B. A. (1994). Effects of gender and sexual orientation on evolutionarily relevant aspects of human mating psychology. *Journal of Personality and Social Psychology, 66*(6), 1081-1093.

Bailey, J. M., Nothnagel, J., & Wolfe, M. (1995). Retrospectively measured individual differences in childhood sex-typed behavior among gay men: Correspondence between self- and maternal reports. *Archives of Sexual Behavior, 24*(6), 613-622.

Baker, S. (1980). Biological influences on sex and gender. *Signs, 6,* 80-96.

Baldwin, J. D., & Baldwin, J .I. (1989). The socialization of homosexuality and heterosexuality in a non-Western society. *Archives of Sexual Behavior, 18*(1), 13-29.

Bancroft, J. (1970). Disorders of sexual potency. In O. Hill (Ed.), *Modern trends in psychosomatic medicine* (pp. 246-259). Norwalk, CT: Appleton & Lang.

Barlow, D. H., & Agras, W. S. (1973). Fading to increase heterosexual responsiveness in homosexuals. *Journal of Applied Behavior Analysis, 6*, 355-366.

Bayer, R. (1987). *Homosexuality and American psychiatry: The politics of diagnosis.* Princeton, New Jersey: Princeton University Press.

Beatty, J. (1995). *Principles of behavioral neuroscience.* Madison, WI: Brown and Benchmark.

Bell, A. P., & Weinberg, M. S. (1978). *Homosexuality: A study of diversity among men and women.* New York: Simon & Schuster.

Bell, A. P., Weinberg, M. S., & Hammersmith, S. K. (1981). *Sexual preference: Its development in men and women.* Bloomington: Indiana University Press.

Bem, D. (1996). Exotic becomes erotic: A developmental theory of sexual orientation. *Psychological Review, 103*(2), 320-335.

Berger, P .L., & Luckmann, T. (1966). *The social construction of reality: A treatise in the sociology of knowledge.* Garden City, NY: Doubleday & Company.

Bergler, E. (1947). Differential diagnosis between spurious homosexuality and perversion homosexuality. *Psychiatric Quarterly, 31*, 399-409.

Bieber, I. (1976). A discussion of "Homosexuality: The ethical challenge." *Journal of Consulting and Clinical Psychology, 44*, 163-166.

Bieber, I., Dain, H. J., Dince, P. R., Drellich, M. G., Grand, H. G., Gundlach, R. H., Kremer, M. W., Rifkin, A. H., Wilber, C. B., & Bieber, T. B. (1962). *Homosexuality: A psychoanalytic study of male homosexuals.* New York: Basic Books.

Birke, L. I. (1982). Is homosexuality hormonally determined? *Journal of Homosexuality, 6*, 35-49.

Blackwood, E. (1993). Breaking the mirror: The construction of lesbianism and the anthropological discourse on homosexuality. In L. D. Garnets & D. C. Kimmel (Eds.), *Psychological perspectives on lesbian and gay male experiences* (pp. 297-315). New York: Columbia University Press.

Bleys, R.C. (1995). *The geography of perversion: Male-to-male sexual behaviour outside of the West and the ethnographic imagination, 1750-1918.* New York: New York University Press.

Blumstein, P., & Schwartz, P. (1976a). Bisexuality in women. *Archives of Sexual Behavior, 5*, 171-181.

Blumstein, P., & Schwartz, P. (1976b). Bisexuality in men. *Urban Life, 5*, 339-358.

Blumstein, P., & Schwartz, P. (1983). *American couples: Money, work, and sex.* New York: William Morrow.

Blumstein, P., & Schwartz, P. (1990). Intimate relationships and the creation of sexuality. In D. P. McWhirter, S. A. Sanders, & J. M. Reinisch (Eds.), *Homosexuality / heterosexuality: concepts of sexual orientation* (pp. 307-320). New York: Oxford University Press.

Boelaars, J. H. M. C. (1950). *The linguistic position of south-western New Guinea.* Leiden, Netherlands: Brill.

Booth, A., & Dabbs, Jr., J. M. (1993). Testosterone and men's marriages. *Social Forces, 72*(2), 463-477.

Booth, A., Shelley, G., Mazur, A., Tharp, G., & Kittok, R. (1989). Testosterone and winning and losing in human competition. *Hormones and Behavior, 23*, 556-571.

Boswell, J. (1980). *Christianity, social tolerance, and homosexuality: Gay people in Western Europe from the beginning of the Christian era to the fourteenth century.* Chicago: University of Chicago Press.

Boswell, J. (1982-1983). Towards the long view: Revolutions, universals and sexual categories. *Salmagundi, 58/59*, 89-113.

Boswell, J. (1990). Categories, experience and sexuality. In E. Stein (Ed.), *Forms of desire: Sexual orientation and the social construction controversy* (pp. 133-173). New York: Routledge.

Boswell, J. (1994). *Same-sex unions in premodern Europe.* New York: Villard Books.

Boxer, A. M., Cook, J. A., & Herdt, G. (1989, August). *First homosexual and heterosexual experiences reported by gay and lesbian youth in an urban community.* Paper presented at the annual meeting of the American Sociological Association, San Francisco, CA.

Boxer, A. M., Cook, J. A., & Herdt, G. (1991). Double jeopardy: Identity transitions and parent-child relations among gay and lesbian youth. In K. Pillemer & K. McCartney (Eds.), *Parent-child relations throughout life* (pp. 59-92). Hillsdale, NJ: Erlbaum.

Breedlove, S. M. (1994). Sexual differentiation of the human nervous system. *Annual Review of Psychology, 45,* 389-418.

Brooten, B. J. (1996). *Love between women: Early Christian responses to female homoeroticism.* Chicago: University of Chicago Press.

Brothers, L. (1997). *Friday's footprints: How society shapes the human mind.* New York: Oxford University Press.

Brown, W. A., Monti, P. M., & Corriveau, D. P. (1978). Serum testosterone and sexual activity and interest in men. *Archives of Sexual Behavior, 7,* 97-104.

Bullough, V. L. (1976). *Sex, society, and history.* New York: Science History Publications.

Bullough, V. L. (1994). *Science in the Bedroom: A History of Sex Research.* New York: Basic Books.

Burns, E. M., Arehart, K. H., & Campbell, S. L. (1992). Prevalence of spontaneous otoacoustic emissions in neonates. *Journal of the Acoustical Society of America, 91,* 1571-1575.

Burr, V. (1995). *An introduction to social constructionism.* London: Routledge.

Buss, D. M. (1998). Sexual strategies theory: Historical origins and current status. *Journal of Sex Research, 35,* 19-31.

Buss, D. M., & Schmitt, D. P. (1993). Sexual strategies theory: An evolutionary perspective on human mating. *Psychological Review, 100,* 204-232.

Buss, D. M., Haselton, M. G., Shakelford, T. K., Bleske, A. L., & Wakefield, J. C. (1998). Adaptations, exaptations, and spandrels. *American Psychologist, 53,* 533-548.

Butler, J. S. (1993, November / December). Homosexuals and the military establishment. *Society, 31,* 13-21.

Byne, W., & Parsons, B. (1993, March). Human sexual orientation: The biologic theories reappraised. *Archives of General Psychiatry, 50,* 228-239.

Campos, P. E., Bernstein, G. S., Davison, G. C., Adams, H. E., & Arias, I. (1996, September). Behavior therapy and homosexuality in the 1990s. *The Behavior Therapist, 19*(8), 113-125.

Cantarella, E. (1992). *Bisexuality in the ancient world.* (C. O'Cuilleanáin, Trans.). New Haven and London: Yale University Press. (Original work published 1988).

Carrier, J. M. (1980). Homosexual behavior in cross-cultural perspective. In J. Marmor (Ed.), *Homosexual behavior: A modern reappraisal* (pp. 100-122). New York: Basic Books.

Christensen, L. W., & Gorski, R. A. (1978). Independent masculinization of neuroendocrine systems by intracerebral implants of testosterone or estradiol in the neonatal female rat. *Brain Research, 146,* 325-340.

Collaer, M. L., & Hines, M. (1995). Human behavioral sex differences: A role for gonadal hormones during early development? *Psychological Bulletin, 118*(1), 55-107.

Cutler, W. B. (1999, January). Human sex-attractant pheromones: Discovery, research, development, and application in sex therapy. *Psychiatric Annals, 29*(1), 54-59.

Cutler, W. B., Friedmann, E., & McCoy, N. L. (1998). Pheromonal influences on sociosexual behavior of men. *Archives of Sexual Behavior, 27,* 1-13.

Dabbs, J. M. (1990). Salivary testosterone measurements: Reliability across hours, days, and weeks. *Physiology & Behavior, 48,* 83-86.

Dalhof, L. G., Hard, E., & Larsson, K. (1977). Influence of maternal stress on offspring sexual behaviour. *Animal Behaviour, 25,* 958-963.

Damon, W. (1983). *Social and personality development.* New York: Norton.

Dannemeyer, W. (1989). *Shadow in the land: Homosexuality in America.* San Francisco: Ignatius.

Darwin, C. (1958). *On the origin of species by means of natural selection.* New York: New American Library. (Original work published 1859).

Darwin, C. (1981). *The descent of man and selection in relation to sex*. Princeton, N J: Princeton University Press. (Original work published 1871).

Davenport, W. H. (1977). Sex in cross-cultural perspective. In F. A. Beach and M. Diamond (Eds.), *Human sexuality in four perspectives* (pp. 155-163). Baltimore: Johns Hopkins University Press.

Dawkins, R. (1996). *Climbing mount improbable*. New York: Norton.

de Lacoste-Utamsing, C., & Holloway, R. L. (1982, June 25). Sexual dimorphism in the human corpus callosum. *Science, 216*, 1431-1432.

de Waal, F. B. M. (1984). Coping with social tension: Sex differences in the effect of food provision to small rhesus monkey groups. *Animal Behavior, 32*, 765-773.

Deacon, A. B. (1934). *Malekula: A vanishing people in the New Hebrides*. London: Routledge.

DeJonge, F. H., Louwerse, A. L., Ooms, M. P., Evers, P., Endert, E., & Van de Poll, N. E. (1989). Lesions of the SDN-POA inhibit sexual behavior of male Wistar rats. *Brain Research Bulletin, 23*, 483-492.

D'Emilio, J., & Freedman, E. B. (1997). *Intimate matters: A history of sexuality in America (2nd ed.)*. Chicago: University of Chicago Press. (Original work published 1988).

Diamant, L., & McAnulty, R .D. (1995). *The psychology of sexual orientation, behavior, and identity: A handbook*. Westport, CT: Greenwood Press.

Dittman, R. W., Kappes M .E., & Kappes, M. H. (1992). Sexual behavior in adolescent and adult females with congenital adrenal hyperplasia. *Psychoneuroendocrinology, 17*, 153-170.

Dobson, J., & Bauer, G. (1990). *Children at risk*. Dallas: Word.

Dörner, G. (1979). Psychoneuroendocrine aspects of brain development and reproduction. In L. Zichella & E. Pancheir (Eds.), *Psychoneuroendocrinology and reproduction*. Amsterdam: Elsevier.

Dörner, G. (1988). Neuroendocrine responses to estrogen and brain differentiation in heterosexuals, homosexuals, and transsexuals. *Archives of Sexual Behavior, 17*, 57-75.

Dörner, G., & Docke, F. (1964). Sex-specific reaction of the hypothalamo-hypophysia system of rats. *Journal of Endocrinology, 30*, 265-266.

Dörner, G., & Hinz, G. (1968). Induction and prevention of male homosexuality by androgens. *Journal of Endocrinology, 40*, 387-388.

Dörner, G., Rhode, W., Stahl, F., *et al.* (1975). A neuroendocrine predisposition for homosexuality in men. *Archives of Sexual Behavior, 4*, 1-8.

Dörner, G., Geier, T., Ahrens, L., *et al.* (1980). Prenatal stress as a possible aetiogenetic factor of homosexuality in human males. *Endokrinologie, 75*, 365-368.

Dörner, G., Poppe, I., Stahl, F., *et al.* (1991). Gene- and environment - dependent neuroendocrine etiogenesis of homosexuality and transsexualism. *Experimental and Clinical Endocrinology, 98*, 141-150.

Dover, K.J. (1989). *Greek homosexuality*. Cambridge, MA: Harvard University Press. (Original work published 1978).

Downey, J., Ehrhardt, A. A., Schiffman, M., Dyrenfurth, I., & Becker, J. (1987). Sex hormones in lesbian and heterosexual women. *Hormones and Behavior, 21*, 347-357.

Dreger, A. D. (1998). *Hermaphrodites and the medical invention of sex*. Cambridge, MA: Harvard University Press.

Dupré, J. (1990). Global versus local perspectives on sexual difference. In D. L. Rhode (Ed.), *Theoretical perspectives on sexual difference* (pp. 47-62). New Haven: Yale University Press.

Dynes, W. R. (1990). Wrestling with the social boa constructor. In E. Stein (Ed.), *Forms of desire: Sexual orientation and the social construction controversy* (pp. 209-238). New York: Routledge.

Ehrhardt, A. A., & Meyer-Bahlburg, H. F. L. (1981). Effects of perinatal hormone on gender-related behavior. *Science, 211*, 1312-1318.

Ehrhardt, A. A., Meyer-Bahlburg, H. F. L., Feldman, J. F., & Ince, S. E. (1984). Sex-dimorphic behavior in childhood subsequent to prenatal exposure to exogenous progestogens and estrogens. *Archives of Sexual Behavior, 13*, 457-477.

Ehrhardt, A. A., Meyer-Bahlburg, H. F. L., Rosen, L .R., Feldman, J. F., Veridiano, N. P., Zimmerman, I., & McEwen, B. S. (1985). Sexual orientation after prenatal exposure to exogenous estrogen. *Archives of Sexual Behavior, 14*, 57-78.

Eliason, M. J. (1995). Accounts of sexual identity formation in heterosexual students. *Sex Roles, 32*(11/12), 821-833.

Ellis, H. (1936). *Studies in the Psychology of Sex.* New York: Random House. (Original work published 1905).

Ellis, L., & Ames, M. A. (1987). Neurohormonal functioning and sexual orientation: A theory of homosexuality-heterosexuality. *Psychological Bulletin, 101*(2), 233-258.

Epstein, S. (1987, May - August). Gay politics, ethnic identity: The limits of social constructionism. *Socialist Review, 93/94*, 9-54.

Faderman, L. (1998). *Surpassing the love of men: Romantic friendship and love between women from the Renaissance to the present.* New York: Quill / William Morrow. (Original work published 1981).

Faludi, S. (1991). *Backlash: The undeclared war against American women.* New York: Crown Publishers, Inc.

Fausto-Sterling, A. (1992). *Myths of gender (2nd ed.).* New York: Basic Books.

Fay, R. E., Turner, C. F., Klassen, A. D., & Gagnon, J. H. (1989). Prevalence and patterns of same-gender sexual contact among men. *Science, 243*, 338-348.

Feldman, M. P., & MacCulloch, M. J. (1971). *Homosexual behavior: Therapy and assessment.* Oxford: Pergamon Press.

Fisher, H. (1998). Lust, attraction, and attachment in mammalian reproduction. *Human Nature, 9*(1), 23-52.

Fisher, S., & Greenberg, R. P. (1977). *The scientific credibility of Freud's theories and therapy.* New York: Basic Books.

Forster, E. M. (1971). *Maurice.* New York: Norton.

Foucault, M. (1990). *The history of sexuality. Vol. I: An introduction.* (R. Hurley, Trans.). New York: Vintage Books. (Original work published 1978).

Fox, R. C. (1995). Bisexual identities. In A. R. D'Augelli & C. J. Patterson (Eds.), *Lesbian, gay, and bisexual identities over the lifespan: Psychological perspectives* (pp. 48-86). New York: Oxford University Press.

Freud, S. (1953). Fragment of an analysis of a case of hysteria. In J. Strachey (Ed.), *The standard edition of the complete psychological works of Sigmund Freud* (Vol. 7; pp. 3-122). London: Hogarth Press. (Original work published 1905a).

Freud, S. (1953). Three essays on the theory of sexuality. In J. Strachey (Ed.), *The standard edition of the complete psychological works of Sigmund Freud* (Vol. 7; pp. 123 - 246). London: Hogarth Press. (Original work published 1905b).

Freud, S. (1953). Analysis of a phobia in a five-year old boy. In J. Strachey (Ed.), *The standard edition of the complete psychological works of Sigmund Freud* (Vol. 10; pp. 1-147). London: Hogarth Press. (Original work published 1909).

Freud, S. (1953). Leonardo da Vinci and a memory of his childhood. In J. Strachey (Ed.), *The standard edition of the complete psychological works of Sigmund Freud* (Vol. 11; pp. 59-138). London: Hogarth Press. (Original work published 1910).

Freud, S. (1953). From the history of an infantile neurosis. In J. Strachey (Ed.), *The standard edition of the complete psychological works of Sigmund Freud* (Vol. 17; pp. 3-22). London: Hogarth Press. (Original work published 1918).

Freud, S. (1953). The psychogenesis of a case of homosexuality in a woman. In J. Strachey (Ed.), *The standard edition of the complete psychological works of Sigmund Freud* (Vol. 18; pp. 155-172). London: Hogarth Press. (Original work published 1920).

Freud, S. (1953). Certain neurotic mechanisms in jealousy, paranoia and homosexuality. In J. Strachey (Ed.), *The standard edition of the complete psychological works of Sigmund Freud* (Vol. 18; pp. 221-234). London: Hogarth Press. (Original work published 1922).

Freud, S. (1953). Some psychical consequences of the anatomical distinction between the sexes. In J. Strachey (Ed.), *The standard edition of the complete psychological works of Sigmund Freud* (Vol. 19; pp. 243-258. London: Hogarth Press. (Original work published 1925).

Frey, U., & Morris, R. G. M. (1997). Synaptic tagging and long-term potentiation. *Nature, 385*, 533-536.

Friedman, R. C. (1988). *Male homosexuality: A contemporary psychoanalytic perspective.* New Haven, CT: Yale University Press.

Gagnon, J. H. (1990). Gender preference in erotic relations: The Kinsey scale and sexual scripts. In D. P. McWhirter, S. A. Sanders, & J. M. Reinisch (Eds.), *Homosexuality / heterosexuality: Concepts of sexual orientation* (pp. 177-207). New York: Oxford University Press.

Gagnon, J. H., & Simon, W. (1973). *Sexual conduct.* Chicago: Aldine.

Gallagher, M., & Chiba, A. A. (1996). The amygdala and emotion. *Current Opinion in Neurobiology, 6*(2), 221-227.

Gallup, G. G., & Suarez, S. D. (1983). Homosexuality as a by-product of selection for optimal heterosexual strategies. *Perspectives in Biology & Medicine, 26*, 315-322.

Gangestad, S. W., & Thornhill, R. (1997). Human sexual selection and developmental stability. In J. A. Simpson & D. T. Kenrick (Eds.), *Evolutionary social psychology* (pp. 169 - 195). Mahwah, NJ: Erlbaum.

Garcia, J., & Koelling, R. (1966). Relation of cue to consequence in avoidance learning. *Psychonomic Science, 4,* 123-124.

Gay, J. (1986). "Mummies and babies" and friends and lovers in Lesotho. *Journal of Homosexuality, 11*(3/4), 55-68.

Geary, D. C. (1998). *Male, female: The evolution of human sex differences.* Washington, D C: American Psychological Association.

Geer, J. H., & Manguno-Mire, G. M. (1996). Gender differences in cognitive processes in sexuality. *Annual Review of Sex Research, 7*, 90-124.

Gergen, K. J. (1985). The social constructionist movement in modern psychology. *American Psychologist, 40*, 266-275.

Geschwind, N., & Galaburda, A. M. (1985). Cerebral lateralization. Part II. *Archives of Neurology, 42*, 521-552.

Gettone, E. (1990). Sappho. In W. R. Dynes (Ed.), *Encyclopedia of homosexuality.* Chicago: St. James Press.

Gladue, B. A., & Bailey, J. M. (1995). Aggressiveness, competitiveness, and human sexual orientation. *Psychoneuroendocrinology, 20*(5), 475-485.

Gladue, B. A., Beatty, W. W., Larson, J., & Staton, R. D. (1990). Sexual orientation and spatial ability in men and women. *Psychobiology, 18*(1), 101-108.

Gladue, B. A., Green, R., & Hellman, R. E. (1984). Neuroendocrine response to estrogen and sexual orientation. *Science, 225*, 1496-1499.

Goldman, R., & Goldman, J. (1982). *Childrens' sexual thinking.* London: Routledge.

Gooren, L. J. G. (1986). The neuroendocrine response of luteinizing hormone to estrogen administration in heterosexual, homosexual and transsexual subjects. *Journal of Clinical Endocrinology and Metabolism, 63*, 583-588.

Gooren, L. J. G. (1995). Biomedical concepts of homosexuality: Folk belief in a white coat. *Journal of Homosexuality, 28*, 237-246.

Gooren, L. J. G., & Cohen-Kettenis, P. T. (1991). Development of male gender identity / role and a sexual orientation towards women in a 46, XY subject with an incomplete form of the androgen insensitivity syndrome. *Archives of Sexual Behavior, 20*(5), 459-470.

Gorski, R. A., Gordon, J. H., Shryne, J. E., & Southam, A. M. (1978). Evidence for a sex difference in the medial preoptic area of the rat brain. *Brain Research, 148*, 333-346.

Gould, S. J. (1991). Exaptation: A crucial tool for evolutionary psychology. *Journal of Social Issues, 47*, 43-65.

Gould, S .J. (1997). The exaptive excellence of spandrels as a term and prototype. *Proceedings of the National Academy of Sciences, 94*, 10750-10755.

Goy, R. W., & McEwen, B .S. (1980). *Sexual differentiation of the brain*. Cambridge, M A: MIT Press.

Graham, S. (1848). *A lecture to young men, on chastity, intended also for the serious consideration of parents and guardians* (10th ed.). Boston: C. H. Pierce.

Green, R. (1985). Gender identity in childhood and later sexual orientation: Follow - up of seventy-eight males. *The American Journal of Psychiatry, 142*, 339-341.

Green, R. (1987). *The "sissy boy syndrome" and the development of homosexuality*. New Haven, CT: Yale University Press.

Greenberg, D. F. (1988). *The construction of homosexuality*. Chicago: University of Chicago Press.

Greenspoon, J., & Lamal, P. A. (1987). A behavioristic approach. In L Diamant (Ed.), *Male and female homosexuality: Psychological approaches* (pp. 109-128). New York: Hemisphere Publishing.

Hacking, I. (1986). Making up people. In T. C. Heller, M. Sosna, & D. E. Wellbery (Eds.), *Reconstructing individualism: Autonomy, individuality, and self in Western thought* (pp. 222-236). Stanford, CA: Stanford University Press.

Haldeman, D. C. (1991). Sexual orientation conversion therapy: A scientific examination. In J. Gonsiorek & J. Weinrich (Eds.), *Homosexuality: Research implications for public policy* (pp. 149-160). Newbury Park, CA: Sage.

Haldeman, D. C. (1994). The practice and ethics of sexual orientation conversion therapy. *Journal of Consulting and Clinical Psychology, 62*(2), 221-227.

Halperin, D. M. (1990). *One hundred years of homosexuality and other essays on Greek love*. New York: Routledge.

Halperin, D. M. (1995). *Saint Foucault: Towards a gay hagiography*. New York: Oxford University Press.

Halpern, D. F. (1986). *Sex differences in cognitive abilities*. Hillsdale, NJ: Erlbaum.

Hamer, D. H., & Copeland, P. (1994). *The science of desire: The search for the gay gene and the biology of behavior*. New York: Simon & Schuster.

Hamer, D. H., Hu, S., Magnuson, V., Hu, N., & Pattatucci, A. M. L. (1993, December 24). Response to male sexual orientation and genetic evidence. *Science, 262*, 2065.

Hamilton, W. D. (1964). The genetical evolution of social behavior. *Journal of Theoretical Biology, 7*, 1-52.

Harasty, J., Double, K. L., Halliday, G. M., Kril, J. J., & McRitchie, D. A. (1997). Language - associated cortical regions are proportionally larger in the female brain. *Archives of Neurology, 54*, 171-176.

Harding, C. F. (1986). The role of androgen metabolism in the activation of male behavior. *The Annals of the New York Academy of Sciences, 474*, 371-378.

Harlow, H. F., & Harlow, M. K. (1965). The affectional systems. In A. M. Schrier, H. F. Harlow & F. Stollnitz (Eds.), *Behavior of nonhuman primates: Modern research trends* (pp. 287-334). New York: Academic Press.

Hendricks, S. E., Graber, B., Rodriguez - Sierra, J. F. (1989). Neuroendocrine responses to exogenous estrogen: No differences between heterosexual and homosexual men. *Psychoneuroendocrinology, 14*(3), 177-185.

Herdt, G. H. (1981). *Guardians of the flutes: Idioms of masculinity*. New York: McGraw-Hill.

Herdt, G. H. (1982). Fetish and fantasy in Sambia initiation. In G. H. Herdt (Ed.), *Rituals of manhood: Male initiation in Papua New Guinea* (pp. 44 - 98). Berkeley : University of California Press.

Herdt, G. H. (1987). *The Sambia: Ritual and gender in New Guinea*. New York: Holt, Rinehart & Winston.

Herdt, G. H. (1990). Mistaken gender: 5 alpha - reductase hermaphroditism and biological reductionism in sexual identity reconsidered. *American Anthropologist, 92*, 433-446.

Herdt, G. H. (1991). Representations of homosexuality in traditional societies: An essay on cultural ontology and historical comparison, part II. *Journal of the History of Sexuality, 2*, 603-632.

Herdt, G. H. (Ed.) (1993). *Ritualized homosexuality in Melanesia.* Berkeley: University of California Press. (Original work published 1984a).

Herdt, G. H. (1993). Semen transactions in Sambia culture. In G. H. Herdt (Ed.), *Ritualized homosexuality in Melanesia* (pp.167-210). Berkeley: University of California Press. (Original work published 1984b).

Herdt, G. H., & Stoller, R. J. (1989). *Intimate communications : Erotics and the study of culture.* New York: Columbia University Press.

Herman, D. (1997). *The antigay agenda: Orthodox vision and the Christian Right.* Chicago: University of Chicago Press.

Hewitt, C. (1995). The socioeconomic position of gay men — A review of the evidence. *American Journal of Economics and Sociology, 54,* 461-479.

Hite, S. (1976). *The Hite report : A nationwide study of female sexuality.* New York: MacMillan.

Hite, S. (1981). *The Hite report on male sexuality.* New York: Knopf.

Hodson, A. (1992). *Essential genetics.* London: Bloomsbury.

Hofman, M. A., & Swaab, D. F. (1989). The sexually dimorphic nucleus of the preoptic area in the human brain: A comparative morphometric study. *Journal of Anatomy, 164,* 55-72.

Hooker, E. (1957). The adjustment of the male overt homosexual. *Journal of Projective Techniques, 21,* 18-31.

Hooper, C. (1992). Biology, brain architecture and human sexuality. *Journal of NIH Research, 4,* 53-59.

Hopkins, P. D. (1992). Gender treachery: Homophobia, masculinity, and threatened identities. In L. May & R. Strikwerda with P. D. Hopkins (Eds.), *Rethinking masculinity: Philosophical explorations in light of feminism* (pp. 111-134). Lanham, MD: Littlefield Adams.

Hoult, T. F. (1984). Human sexuality in biological perspective: Theoretical and methodological considerations. *Journal of Homosexuality, 9,* 137-155.

Hu, S., Pattatucci, A. M. L., Patterson, C., Li., L., Fulker, D. W., Cherny, S. S., Kruglyak, L., & Hamer, D. H. (1995). Linkage between sexual orientation and chromosome Xq28 in males but not in females. *Nature Genetics, 11,* 248-256.

Hubbard, R., & Wald, E. (1993). *Exploding the gene myth.* Boston: Beacon Press.

Hutchison, J. B., & Beyer, C. (1994). Gender - specific brain formation of oestrogen in behavioural development. *Psychoneuroendocrinology, 19*(5-7), 529-541.

Imperato-McGinley, J., Guerrero, L., Gautier, T., & Peterson, R. E. (1974). Steroid 5 alpha - reductase deficiency in man: An inherited form of male pseudohermaphroditism. *Science, 186,* 1213-1215.

Imperato-McGinley, J., Peterson, R. E., Gautier, T., & Sturla, E. (1979, May). Androgens and the evolution of male - gender identity among male pseudohermaphrodites with 5 alpha - reductase deficiency. *New England Journal of Medicine, 300,* 1233-1237.

Imperato-McGinley, J., Miller, M., Wilson, J. D., Peterson, R. E., Shackleton, C., & Gajdusek, D.C. (1991). A cluster of male pseudo-hermaphrodites with 5 alpha - reductase deficiency in Papua New Guinea. *Clinical Endocrinology, 34,* 293-298.

Isay, R. (1989). *Being homosexual: Gay men and their development.* New York: Farrar, Straus, Giroux.

Janus, S. S., & Janus, C. L. (1993). *The Janus report on sexual behavior.* New York: Wiley.

Johnson, L., George, F. W., Neaves, W. B., Rosenthal, I. M., Christensen, R. A., Decristoforo, A., Schweikert, H., Sauer, M. V., Leshin, M., Griffin, J. E., & Wilson, J. D. (1986). Characterization of the testicular abnormality in 5 alpha - reductase deficiency. *Journal of Clinical Endocrinology Metabolism, 63,* 1091-1099.

Johnson, A. M., Wadsworth, J., Wellings, K., Bradshaw, S., & Field, J. (1992). Sexual lifestyles and HIV risk. *Nature, 360,* 410-412.

Johnson, A. M., Wadsworth, J., Wellings, K., & Field, J. (1994). *Sexual attitudes and lifestyles.* Oxford: Blackwell Scientific.

Kallmann, F. J. (1952). Comparative twin study on the genetic aspects of male homosexuality. *Journal of Nervous and Mental Disease, 115,* 283-298.

Katz, J. N. (1995). *The invention of heterosexuality.* New York: Dutton.

Kauth, M. R., & Landis, D. (1996). Applying lessons learned from minority integration in the military. In G. M. Herek, J. B. Jobe, & R. M. Carney (Eds.), *Out in force: Sexual orientation and the military* (pp. 86-105). Chicago: University of Chicago Press.

Kauth, M. R., & Kalichman, S. C. (1995). Sexual orientation and development: An interactive approach. In L. Diamant & R. D. McAnulty (Eds.), *The psychology of sexual orientation, behavior, and identity: A handbook* (pp. 81-103). Westport, CT: Greenwood Press.

Kawata, M., Yuri, K., & Morimoto, M. (1994). Steroid hormone effects on gene expression, neuronal structure, and differentiation. *Hormones and Behavior, 28,* 477-482.

Kellogg, J. H. (1974). *Plain facts for old and young.* Buffalo, NY: Heritage Press. (Original work published 1882).

Kester, P., Green, R., Finch, S. J., & Williams, K. (1980). Prenatal " female hormone " administration and psychosexual development in human males. *Psychoneuroendocrinology, 5,* 269-285.

Kinsey, A. C., Pomeroy, W. B., & Martin, C. E. (1948). *Sexual behavior in the human male.* Philadelphia: W. B. Saunders.

Kinsey, A. C., Pomeroy, W. B., Martin, C. E., & Gebhard, P. H. (1953). *Sexual behavior in the human female.* Philadelphia: W. B. Saunders.

Kirsch, J. A. W., & Weinrich, J. D. (1991). Homosexuality , nature , and biology : Is homosexuality natural ? Does it matter ? In J. C. Gonsiorek & J. D. Weinrich (Eds.), *Homosexuality: Research implications for public policy* (pp. 13 - 31). Newbury Park, CA: SAGE.

Klein, F., Sepekoff, B., & Wolf, T. J. (1985). Sexual orientation: A multi - variate dynamic process. *Journal of Homosexuality, 11*(1/2), 35-49.

Knauft, B. M. (1985). *Good company and violence: Sorcery and social action in a lowland New Guinea society.* Berkeley: University of California Press.

Knauft, B. M. (1987). Homosexuality in Melanesia . *The Journal of Psychoanalytic Anthropology, 10*(2), 155-191.

Knussman, R., Christiansen, K., & Couwenberge, C. (1986). Relationships between sex hormone levels and sexual behavior in men. *Archives of Sexual Behavior, 15,* 429-445.

Kuhn, T. (1996). *The structure of scientific revolutions.* Chicago: University of Chicago Press. (Original work published 1962).

Kurdek, L. A. (1995). Lesbian and gay couples. In A. R. D'Augelli & C. J. Patterson (Eds.), *Lesbian, gay, and bisexual identities over the lifespan: Psychological perspectives* (pp. 243 - 261). New York: Oxford.

Lancaster, R. N. (1995). That we should all turn queer? Homosexual stigma in the making of manhood and the breaking of a revolution in Nicaragua. In R. G. Parker & J. H. Gagnon (Eds.), *Conceiving sexuality: Approaches to sex research in a postmodern world* (pp. 135 - 156). New York: Routledge.

Langevin, R., & Martin, M. (1975). Can erotic responses be classically conditioned? *Behavior Therapy, 6,* 350-355.

Laumann, E., Gagnon, J., Michael, R., & Michaels, S. (1994). *The social organization of sexuality: Sexual practices in the United States.* Chicago: University of Chicago Press.

Leupp, G. P. (1995). *Male colors: The construction of homosexuality in Tokugawa Japan.* Berkeley, CA: University of California Press.

LeVay, S. (1991). A difference in hypothalamic structure between heterosexual and homosexual men. *Science, 253,* 1034-1037.

LeVay, S. (1993). *The sexual brain.* Cambridge, MA: MIT Press.

Lewes, K. (1988). *The psychoanalytic theory of male homosexuality.* New York: Meridian.

Lindenbaum, S. (1987). The mystification of female labors. In J. F. Collier & S. J. Yangisako (Eds.), *Gender and kinship* (pp. 221-243). Stanford, CA: Stanford University Press.

MacLusky, N. J., & Naftolin, F. (1981, March 20). Sexual differentiation of the central nervous system. *Science, 211,* 1294-1303.

Malyon, A. K. (1981). The homosexual adolescent: Developmental issues and social bias. *Child Welfare, 60,* 321-330.

Mange, A. P., & Mange, E. J. (1990). *Genetics: Human aspects, second edition.* Sunderland, MA: Sinauer Associates.

Manguno-Mire, G. M, & Geer, J. H. (1998). Network knowledge organization: Do knowledge structures for sexual and emotional information reflect gender or sexual orientation? *Sex Roles, 39*(9/10), 705-529.

Marks, I. M., & Gelder, M. G. (1967). Transvestism and fetishism: Clinical and psychological changes during faradic aversion. *British Journal of Psychiatry, 113,* 711-729.

Marshall, J. (1981). Pansies, perverts and macho men : Changing conceptions of male homosexuality. In K. Plummer (Ed.), *The making of the modern homosexual* (pp. 133 - 154). Totowa, NJ: Barnes & Noble Books.

Martin, R. K. (1989). Knights-errant and gothic seducers: The representation of male friendship in mid-nineteenth-century America. In M. B. Duberman, M. Vicinus, & G. Chauncey, Jr. (Eds.), *Hidden from history: Reclaiming the gay and lesbian past* (pp. 169-182). New York: NAL Books.

Masters, W. H., & Johnson, V. E. (1979). *Homosexuality in perspective.* Boston: Little, Brown.

McClintock, M. K. (1971). Menstrual synchrony and suppression. *Nature, 291,* 244-245.

McClintock, M. K. (1984). Estrous synchrony: Modulation of ovarian cycle length by female pheromones. *Physiological Behavior, 32,* 701-705.

McConaghy, N. (1967). Penile volume change to moving pictures of male and female nudes in heterosexual and homosexual males. *Behavior Research and Therapy, 5,* 43-48.

McConaghy, N. (1987). A learning approach. In J. H. Geer & W. T. O' Donoghue (Eds.), *Theories of human sexuality* (pp. 287-333). New York: Plenum.

McConaghy, N. (1993). *Sexual behavior: Problems and management.* New York: Plenum.

McConaghy, N., Buhrich, N., & Silove, D. (1994). Opposite sex - linked behaviors and homosexual feelings in the predominantly heterosexual male majority. *Archives of Sexual Behavior, 23*(5), 565-577.

McCormick, C. M., & Witelson, S. F. (1991). A cognitive profile of homosexual men compared to heterosexual men and women. *Psychoneuroendocrinology, 16*(6), 459-473.

McCormick, C. M., Witelson, S. F., & Kingstone, E. (1990). Left - handedness in homosexual men and women: Neuroendocrine implications. *Psychoneuroendocrinology, 15,* 69-76.

McFadden, D. (1998). Sex differences in the auditory system. *Developmental Neuropsychology, 14*(2/3), 261-298.

McFadden, D., Loehlin, J. C., & Pasanen, E. G. (1996). Additional findings on heritability and prenatal masculinization of cochlear mechanisms : Click - evoked otoacoustic emissions. *Hearing Research, 97,* 102-119.

McFadden, D., & Pasanen, E. G. (1998, March). Comparison of the auditory systems of heterosexuals and homosexuals: Click - evoked otoacoustic emissions. *Proceedings of the National Academy of Sciences, 95,* 2709-2713.

McFadden, D., Pasanen, E. G., & Callaway, N. L. (1998, September). Changes in otoacoustic emissions in a transsexual male during treatment with estrogen. *Journal of the Acoustical Society of America, 104*(3), 1555-1558.

McFadden, D., & Pasanen, E. G. (1999, April). Spontaneous otoacoustic emissions in heterosexuals, homosexuals, and bisexuals. *Journal of the Acoustical Society of America, 105*(4), 2403-2413.

McGraw, K. O. (1987). *Developmental psychology.* Orlando, FL: Harcourt Brace Jovanovich.

McGuire, T. (1995). Is homosexuality genetic? A critical review and some suggestions. *Journal of Homosexuality, 28*(1/2), 115-145.

McHenry, H. M. (1994). Behavioral ecological implications of early hominid body size. *Journal of Human Evolution, 27,* 77-87.

McIntosh, M. (1990). The homosexual role. In E. Stein (Ed.), *Forms of desire : Sexual orientation and the social constructionist controversy* (pp. 25 - 42). New York: Routledge. (Original work published 1968).

McKernan, M. G., & Shinnick-Gallagher, P. (1997, December). Fear conditioning induces a lasting potentiation of synaptic currents in vitro. *Nature, 390,* 607-611.

McKnight, J. (1997). *Straight science ? Homosexuality, evolution and adaptation.* London: Routledge.

McMullen, R. (1982). Roman attitudes to Greek love. *Historia, 31,* 484-502.

McWhirter, D. P., & Mattison, A. M. (1984). *The male couple.* Englewood Cliffs, NJ: Prentice-Hall.

Meyer-Bahlburg, H. F. L. (1984). Psychoendocrine research on sexual orientation: Current status and future options. *Progress in Brain Research, 61,* 375-398.

Money, J. (1969). Sexually dimorphic behavior, normal and abnormal. In N. Kretchmer & D.N. Walcher (Eds.), *Environmental influences on genetic expression.* Washington, DC: U. S. Government Printing Office.

Money, J. (1970). Sexual dimorphism and homosexual gender identity. *Psychological Bulletin, 74,* 425-440.

Money, J. (1988). *Gay, straight, and in - between: The sexology of erotic orientation.* New York: Oxford University Press.

Money, J. (1998). *Sin, science, and the sex police.* New York: Prometheus Books.

Money, J., & Daléry, J. (1977). Hyperadrenocortical 46, XX hermaphroditism with penile urethra: Psychological studies in seven cases, three reared as boys, four as girls. In P. A. Lee, L P. Plotnick, A. A. Kowarski, & C. J. Migeon (Eds.), *Congential adrenal hyperplasia* (pp. 433-446). Baltimore: University Park Press.

Money, J., & Ehrhardt, A. A. (1972). *Man and woman, boy and girl: The differentiation and dimorphism of gender identity from conception to maturity.* Baltimore: Johns Hopkins Press.

Money, J., Schwarz, M., & Lewis, V. G. (1984). Adult heterosexual status and fetal hormonal masculinization and demasculinization: 46, XX congenital virilizing adrenal hyperplasia and 46, XY androgen insensitivity syndrome compared. *Psychoneuroendocrinology, 9,* 405-414.

Moore, C. L. (1986). Interaction of species - typical environmental and hormonal factors in sexual differentiation of behavior. *The Annals of the New York Academy of Sciences, 474,* 108-119.

Murphy, T. F. (1992). Redirecting sexual orientation : Techniques and justifications. *The Journal of Sex Research, 29*(4), 501-523.

Nass, R. Baker, S., Speiser, P., Virdis, R., Balmaso, A., Cacciari, E., Loche, A., Dumic, M., & New, M. (1987). Hormones and handedness: Left - hand bias in female congenital adrenal hyperplasia patients. *Neurology, 37,* 711-715.

Netley, C., & Rovet, J. (1982). Atypical hemispheric lateralization in Turner syndrome subjects. *Cortex, 18,* 377-384.

Nicolosi, J. (1991). *Reparative therapy of male homosexuality : A new clinical approach.* Northvale, NJ: Jason Aronson.

Nordeen, E. J., Yahr, P. (1983). A regional analysis of estrogen binding to hypothalamic cell nuclei in relation to masculinzation and defeminization. *Journal of Neuroscience, 3,* 933-941.

Norton, R. (1992). *Mother clap's molly house: The gay subculture in England 1700 - 1830.* London: Gay Men's Press.

Norton, R. (1997). *The myth of the modern homosexual: Queer history and the search for cultural unity.* London: Cassell.

"Nunn offers a compromise: 'Don't ask, don't tell.'" (1993, May 15). *Congressional Quarterly Weekly Report,* pp. 1240-1242.

Ogawa, S., Lubahn, D. B., Kurach, K. S., & Pfaff, D. W. (1997). Behavioral effects of estrogen receptor gene disruption in male mice. *Proceedings of the National Academy of Sciences, 94,* 1476-1481.

Oosterhuis, H. (1991). Male bonding and homosexuality in German nationalism. In H. Oosterhuis & H. Kennedy (Eds.), *Homosexuality and male bonding in pre-nazi Germany* (pp. 241-264). New York: Harrington Park Press.

Padgug, R. (1990). Sexual matters: On conceptualizing sexuality in history. In E. Stein (Ed.), *Forms of desire: Sexual orientation and the social constructionist controversy* (pp. 43 - 67). New York: Routledge. (Original work published 1979).

Pattatucci, A. M. L., & Hamer, D. (1995). Development and familiality of sexual orientation in females. *Behavioral Genetics, 25,* 4107-4120.

Paulk, J. (1998). *Not afraid to change: The remarkable story of how one man overcame homosexuality.* Yakima, WA: WinePress Publishing.

Pennington, S. (1989). *Ex-gays? There are none.* Hawthorne: Lambda Christian Fellowship.

Pfaff, D. W. (1980). *Estrogens and brain function: Neural analysis of a hormone - controlled mammalian reproductive behavior.* New York: Springer-Verlag.

Plummer, K. (Ed.) (1981). *The making of the modern homosexual.* Totowa, N J : Barnes & Noble Books.

Pritchard, J. B. (1958). *The ancient Near East.* Princeton, NJ: Princeton University Press.

Pronger, B. (1990). *The arena of masculinity: Sports, homosexuality, and the meaning of sex.* New York: St. Martin's Press.

Rachman, S. (1966). Sexual fetishism: An experimental analogue. *The Psychological Record, 16,* 293-296.

Rachman, S., & Hodgson, R. J. (1968). Experimentally-induced "sexual fetishism": Replication and development. *The Psychological Record, 18,* 25-27.

Rado, S. (1940). A critical examination of the concept of bisexuality. *Psychosomatic Medicine, 2,* 459-467.

Reinisch, J. M., & Sanders, S. A. (1992). Effects of prenatal exposure to diethlstilbestrol (DES) on hemispheric laterality and spatial ability in human males. *Hormones and Behavior, 26,* 62-75.

Reiss, I. (1986). *Journey into sexuality: An exploratory voyage.* Englewood Cliffs, N J: Prentice-Hall.

Reite, M., Sheeder, J., Richardson, D., & Teale, P. (1995). Cerebral laterality in homosexual males: Preliminary communication using magnetoencephalography. *Archives of Sexual Behavior, 24,* 585-593.

Remafedi, G., Resnick, M., Blum, R., & Harris, L. (1992). Demography of sexual orientation in adolescents. *Pediatrics, 89,* 714-721.

Rice, G., Anderson, C., Risch, N., & Ebers, G. (1999, April 23). Male homosexuality: Absence of linkage to microsatellite markers at Xq28. *Science, 284,* 665-667.

Ridley, M. (1994). *The red queen: Sex and the evolution of human nature.* New York: Penguin Books.

Risch, N., Squires-Wheeler, E., & Keats, B. J. B. (1993, December 24). Male sexual orientation and genetic evidence. *Science, 262,* 2063-2064.

Robbins, T. W., & Everitt, B. J., (1996). Neurobehavioural mechanisms of reward and motivation. *Current Opinion in Neurobiology, 6*(2), 228-236.

Rodriguez-Sierra, J. F. (1986). Extended organizational effects of estrogen at puberty. *The Annals of the New York Academy of Sciences, 474,* 293-307.

Rogan, M. T., Staeubil, U. V., & LeDoux, J. E. (1997). Fear conditioning induces associative long-term potentiation in the amygdala. *Nature, 390,* 604-607.

Roheim, G. (1933). Women and their life in Central Australia. *Journal of the Royal Anthropological Institute of Great Britain and Ireland, 63,* 207-265.

Ross, M. W., & Arrindell, W. A. (1988). Perceived parental rearing practices of homosexual and heterosexual men. *Journal of Sex Research, 24,* 275-281.

Ruffie, J. (1986). *The population alternative: A new look at competition and the species.* New York: Penguin Books.

Russo, V. (1981). *The celluloid closet: Homosexuality in the movies.* (Rev. ed.). New York: Harper & Row.

Sandberg, D. E., Meyer-Bahlburg, H. F. L., Yager, T. J., Hensle, T. W., Levitt, S. B., Kogan, S. J., & Reda, E. F. (1995). Gender development in boys born with hypospadias. *Psychoneuroendocrinology, 20*(7), 693-709.

Sanders, S. A., & Reinisch, J. M. (1999, January 20). Would you say you "had sex" if ...? *Journal of the American Medical Association, 281*(3), 275-277.

Sapir, E. (1947). *Selected writings in language, culture, and personality.* Los Angeles: University of California Press.

Sarbin, T. R., & Karols, K. E. (1988). *Nonconforming sexual orientations and military suitability*. PERS-TR-89-002. Defense Personnel Security Research and Education Center.

Scher, S. J. (1999). Are adaptations necessarily genetic? *American Psychologist, 54*(6), 436.

Schüklenk, U., & Ristow, M. (1996). The ethics of research into the cause(s) of homosexuality. *Journal of Homosexuality, 31*(3), 5-30.

Schwimmer, E. (1993). Male couples in New Guinea. In G. H. Herdt (Ed.), *Ritualized Homosexuality in Melanesia*, (pp. 248 - 291). Berkeley: University of California Press. (Original work published in 1984).

Seligman, M. E. P. (1970). On the generality of the laws of learning. *Psychological Review, 77*, 406-418.

Serpenti, L. M. (1965). *Cultivators in the swamps: Social structure and horticulture in a New Guinea society*. Assen: Van Gorcum.

Serpenti, L. M. (1993). The ritual meaning of homosexuality and pedophilia among the Kimam-Papuans of South Irian Jaya. In G. H. Herdt (Ed.), *Ritualized homosexuality in Melanesia*, (pp. 292-317). Berkeley: University of California Press. (Original work published 1984).

Shepherd, G. (1987). Rank, gender, and homosexuality: Mombasa as a key to understanding sexual options. In P. Caplan (Ed.), *The cultural construction of sexuality*, (pp. 240 - 270). London: Tavistock Publishers.

Signorella, M. L., & Jamison, W. (1986). Masculinity, femininity, androgyny and cognitive performance: A meta-analysis. *Psychological Bulletin, 100*, 207-228.

Signorile, M. (1993). *Queer in America: Sex, the media, and the closets of power*. New York: Random House.

Sloan, I. J. (1987). *Homosexual conduct and the law*. London: Oceana Publications.

Socarides, C. (1968). The overt homosexual. New York: Grune & Stratton.

Socarides, C. (1969). Psychoanalytic therapy of a male homosexual. *Psychoanalytic Quarterly, 38*, 173-190.

Socarides, C., & Volkan, V. (1990). *The homosexualities: Reality, fantasy, and the arts*. Madison, CT: International University Press.

Sørum, A. (1993). Growth and decay: Bedamini notions of sexuality. In G. H. Herdt (Ed.), *Ritualized homosexuality in Melanesia*, (pp. 318 - 336). Berkeley: University of California Press. (Original work published 1984).

Starr, K. (1998, November 17). *The Starr report: The official report of the independent counsel's investigation of the President*. http://www.thestarreport.org/

Stein, E. (Ed.). (1990). *Forms of desire: Sexual orientation and the social constructionist controversy*. New York: Routledge.

Stein, E. (1999). *The mismeasure of desire : The science, theory, and ethics of sexual orientation*. New York: Oxford University Press.

Stern, K., & McClintock, M. K. (1998). Regulation of ovulation by human pheromones. *Nature, 392*, 177-179.

Stoller, R. J., & Herdt, G. H. (1985). Theories of origins of male homosexuality: A cross - cultural look. In R. J. Stoller (Ed.), *Observing the erotic imagination*, (pp. 104 - 134). New Haven: Yale University Press.

Swaab, D. F., & Fliers, E. (1985). A sexually dimorphic nucleus in the human brain. *Science, 228*, 1112-1115.

Swaab, D. F., & Hofman, M. A. (1988). Sexual differentiation of the human hypothalamus: Ontogeny of the sexually dimorphic nucleus of the preoptic area. *Developmental Brain Research, 44*, 314-318.

Swaab, D. F., & Hofman, M. A. (1990). An enlarged suprachiasmaticnucleus in homosexual men. *Brain Research, 537*, 141-148.

Swaab, D. F., Gooren, L. J. G., & Hofman, M. A. (1992). Gender and sexual orientation in relation to hypothalamic structures. *Hormone Research, 38*, 51-61.

Symons, D. (1987). *The evolution of human sexuality*. New York: Oxford University Press. (Original work published 1979).

Thornhill, R., & Gangestad, S. W. (1994). Human fluctuating asymmetry and sexual behavior. *Psychological Science, 5*(5), 297-302.

Thornton, J. E. (1986). Heterotypical sexual behavior: Implications from variations. *The Annals of the New York Academy of Sciences, 474*, 362-370.

Trivers, R. L. (1971). The evolution of reciprocal altruism. *Quarterly Review of Biology, 46*(4), 35-57.

Trivers, R. L. (1974). Parent-offspring conflict. *American Zoologist, 14*, 249-264.

Trivers, R. L. (1985). *Social evolution.* Menlo Park, CA: Benjamin/Cummings.

Trumbach, R. (1988). Gender and the homosexual role in modern Western culture: The Eighteenth and nineteenth centuries compared. In D. Altman, C. Vance, M. Vicinus, J. Weeks, *et al.* (Eds.), *Homosexuality, which homosexuality?,* (pp. 149 - 169). London: Gay Mens Press.

Tsitouras, P. D., Martin, C. E., & Harman, S. M. (1982). Relationship of serum testosterone to sexual activity in healthy elderly men. *Journal of Gerontology, 37*, 288-293.

Udry, J. R., Billy, J. O. G., Morris, N. M, Groff, T. R., & Raj, H. R. (1985). Serum androgenic hormones motivate sexual behavior in adolescent boys. *Fertility and Sterility, 43*, 90-94.

van Gulik, R. H. (1974). *Sexual life in ancient China: A preliminary survey of Chinese sex and society from ca. 1500 B.C. till 1644 A.D.* Leiden, The Netherlands: E. J. Brill. (Original work published 1961).

Van Wyck, P. H., & Geist, C. S. (1984). Psychosocial development in heterosexual, bisexual, and homosexual behavior. *Archives of Sexual Behavior, 13*, 505-544.

Vance, C. S. (1991). Anthropology rediscovers sexuality: A theoretical comment. *Social Science Medicine, 33*(8), 875-884.

vom Saal, F. S. (1989). Sexual differentiation in litter-bearing mammals: Influence of sex on adjacent fetuses in utero. *Journal of Animal Science, 67*, 1824-1840.

Wada, J. A., Clarke, R., & Hamm, A. (1975). Cerebral hemispheric asymmetry in humans. *Archives of Neurology, 32*, 58-70.

Wallace, B. (1968). *Topics in population genetics.* New York: Norton.

Wallen, K. (1990). Desire and ability: Hormones and the regulation of female sexual behavior. *Neuroscience & Biobehavioral Review, 14*, 233-241.

Warmington, E. H., & Rouse, P. G. (1984). *Great Dialogues of Plato.* New York: New American Library.

Wedekind, C., Seebeck, T., Bettens, F., & Paepke, A. J. (1995). MHC - dependent mate preferences in humans. *Proceedings of the Royal Society of London, 260*(1359), 245-249.

Weeks, J. (1977). *Coming out: Homosexual politics in Britain from the nineteenth century to the present.* London: Quartet Books.

Weeks, J. (1981). Discourse, desire and sexual deviance: Some problems in a history of homosexuality. In K. Plummer (Ed.), *The making of the modern homosexual,* (pp. 76 - 111). Totowa, NJ: Barnes & Noble Books.

Weeks, J. (1985). *Sexuality and its discontents: Meanings, myths & modern sexualities.* London: Routledge & Kegan Paul.

Weeks, J. (1988). Against nature. In D. Altman, C. Vance, M. Vicinus, J. Weeks, *et al.* (Eds.), *Homosexuality, which homosexuality?,* (pp. 199-213). London: Gay Mens Press.

Weeks, J. (1995). History, desire, and identities. In R. G. Parker & J. H. Gagnon (Eds.), *Conceiving sexuality: Approaches to sex research in a postmodern world,* (pp. 33 - 50). New York: Routledge.

Weinrich, J. D. (1987a). *Sexual landscapes: Why we are what we are, why we love whom we love.* New York: Scribner's.

Weinrich, J. D. (1987b). A new sociobiological theory of homosexuality applicable to societies with universal marriage. *Ethology and Sociobiology, 8*, 37-47.

Weinrich, J. D. (1990). Reality or social constructionism? In E. Stein (Ed.), *Forms of desire: Sexual orientation and the social construction controversy,* (pp. 175 - 208). New York: Routledge.

Whalen, R. E. (1991). Heterotypical behaviour in man and animals: Concepts and strategies. In M. Haug, P.F. Brain, & C. Aron (Eds.), *Heterotypical behaviour in man and animals*, (pp. 215-227). London: Chapman & Hall.

Whitam, F. L., Diamond, M., & Martin, J. (1993). Homosexual orientation in twins: a report on 61 pairs and 3 triplet sets. *Archives of Sexual Behavior, 22,* 187-206.

Wieringa, S. (1988). An anthropological critique of constructionism: Berdaches and butches. In D. Altman, C. Vance, M. Vicinus, J. Weeks, *et al.* (Eds.), *Homosexuality, which homosexuality?*, (pp. 215-238). London: Gay Men's Press.

Williams, C. L. (1986). A reevaluation of the concept of separable periods of organizational and activational actions of estrogens in development of brain and behavior. *The Annals of the New York Academy of Sciences, 474,* 282-292.

Williams, F. E. (1936). *Papuans of the Trans-Fly.* Oxford: Oxford University Press.

Williams, L. (1993, June 28). Blacks reject gay rights fight as equal to theirs. The New York Times, pp. A1, A12.

Williams, W. (1986). *The spirit and the flesh: Sexual diversity in American Indian culture.* Boston: Beacon Press.

Wilson, E. O. (1975). *Sociobiology: The new synthesis.* Cambridge, MA: Harvard University Press.

Wilson, E. O. (1978). *On human nature.* Cambridge, MA: Harvard University Press.

Witt, L., Thomas, S., & Marcus, E. (Eds.). (1995). *Out in all directions: The almanac of gay and lesbian America.* New York: Warner Books.

Wolff, C. (1971). *Love between women.* New York: Harper & Row.

Wolpe, J. (1958). *Psychotherapy by reciprocal inhibition.* Stanford, CA: Stanford University Press.

Zilbergeld, B. (1992). *The new male sexuality: The truth about men, sex, and pleasure.* New York: Bantam Books.

Index